Missiles for the Nineties

Westview Replica Editions

The concept of Westview Replica Editions is a response to the continuing crisis in academic and informational publishing. Library budgets for books have been severely curtailed. Ever larger portions of general library budgets are being diverted from the purchase of books and used for data banks, computers, micromedia, and other methods of information retrieval. Interlibrary loan structures further reduce the edition sizes required to satisfy the needs of the scholarly community. Economic pressures on the university presses and the few private scholarly publishing companies have severely limited the capacity of the industry to properly serve the academic and research communities. As a result, many manuscripts dealing with important subjects, often representing the highest level of scholarship, are no longer economically viable publishing projects--or, if accepted for publication, are typically subject to lead times ranging from one to three years.

Westview Replica Editions are our practical solution to the problem. We accept a manuscript in camera-ready form, typed according to our specifications, and move it immediately into the production process. As always, the selection criteria include the importance of the subject, the work's contribution to scholarship, and its insight, originality of thought, and excellence of exposition. The responsibility for editing and proofreading lies with the author or sponsoring institution. We prepare chapter headings and display pages, file for copyright, and obtain Library of Congress Cataloging in Publication Data. A detailed manual contains simple instructions for preparing the final typescript, and our editorial staff is always available to answer questions.

The end result is a book printed on acid-free paper and bound in sturdy library-quality soft covers. We manufacture these books ourselves using equipment that does not require a lengthy make-ready process and that allows us to publish first editions of 300 to 600 copies and to reprint even smaller quantities as needed. Thus, we can produce Replica Editions quickly and can keep even very specialized books in print as long as there is a demand for them.

About the Book and Editors

Missiles for the Nineties:
ICBMs and Strategic Policy
edited by Barry R. Schneider, Colin S. Gray,
and Keith B. Payne

This book addresses the major issues in the current debate over the building and deployment of silo-based ICBMs in the U.S. The authors consider the role of the MX in deterrence strategy, the impact of new technology on U.S. strategic planning options, likely responses of Soviet military planners, and the effects that new U.S. ICBMs would have on U.S.-Soviet arms control negotiations. They also discuss the implications of President Reagan's Strategic Defense Initiative for future decisions regarding ICBM deployment.

Barry R. Schneider is senior defense analyst at the National Institute for Public Policy and co-editor of Current Issues in U.S. Defense Policy. **Colin S. Gray** is president of the National Institute for Public Policy and author of American Military Space Policy (1983). **Keith B. Payne**, executive vice president and director of national security studies at the National Institute for Public Policy, is author of Nuclear Deterrence in U.S.-Soviet Relations (Westview, 1982) and editor of Laser Weapons in Space (Westview, 1983).

Missiles for the Nineties

ICBMs and Strategic Policy

edited by
Barry R. Schneider,
Colin S. Gray, and
Keith B. Payne

Foreword by
R. James Woolsey

Westview Press / Boulder and London

A Westview Replica Edition

Published in 1984 in the United States of America by
    Westview Press, Inc.
    5500 Central Avenue
    Boulder, Colorado 80301
    Frederick A. Praeger, Publisher

Library of Congress Catalog Card Number: 84-40410
ISBN 0-8133-7001-9

Printed and bound in the United States of America

10 9 8 7 6 5 4 3 2 1

Contents

Tables and Figures

Foreword

R. James Woolsey

For most of its early life the ICBM seemed to be the answer to a strategist's dream: invulnerable in its silo, inexpensive to maintain, promptly available to the President, easily communicated with, comparatively accurate. It mattered little whether you fell on the assured destruction or the warfighting side of what was for many years the Great Divide in the American strategic debate. Whatever the question, the ICBM was the answer -- the modern equivalent of Zeus's lightning bolts, secure in the nation's hands.

By the end of the 1960s, however, continued technical improvements in one of the ICBM's positive features -- accuracy -- and the development of MIRVs began to undermine the ICBM's strategic position. From the point of view of American strategic interests, it would have been positive news -- and more credit to the U.S. ICBM force -- if it were only American ICBM accuracy that was continuing to improve from good to very good and only American MIRVs that were being deployed. But by as early as the beginning of the Nixon Administration in 1969 it was widely thought to be prudent to begin to worry about accurate MIRVed Soviet ICBMs threatening the silo-based U.S. ICBM force. Indeed the concern led to a shift in emphasis for the role of the embryonic U.S. ABM system in 1969. "Sentinel" became "Safeguard" and the principal role for a U.S. ABM system became, in the course of that year's debate, to protect Minuteman against the threat posed by the new large Soviet SS-9 ICBM.

Basing U.S. ICBMs in ordinary silos, unprotected, thus began to have strategic and political problems some fifteen years ago. At least since then, the perennial question has been, and continues to be, how long can silo-basing of ICBMs be a reasonable thing to do? In the absence of the ability to dig relatively inexpensive holes on the surface of the earth, put ICBMs in them, and have a fair degree of confidence in their survivability, the ICBM portion of the strategic equation changes -- especially in a democracy that must deal with these matters in a public debate. There is no longer an easy community of interest between those of different strategic persuasions.

Some have thus looked at ICBMs in isolation from the rest of our strategic forces and have ignored the complications that the Soviets would face -- with today's systems -- in attacking our ICBMs and our bombers simultaneously. Many have also explicitly or implicitly minimized the role of ICBMs as part of NATO's nuclear shield against a massive conventional Soviet attack. Both such views have helped produce an early disenchantment with ICBMs in silos and, for many, disenchantment with ICBMs -- period.

Others have labored hard to find early and simple solutions to the problem of survivability. But no quick or semi-quick fix with current systems, including MX, has been as satisfactory as was the ICBM system in its prime -- when, sitting in silos, it was cheap to maintain, promptly available, accurate, and, above all, survivable.

Thus nostalgia has, for a time, become the order of the day. At its worst this has produced the phenomenon of denial -- denial that Soviet missiles can ever achieve a high degree of accuracy, denial that we need to be concerned about it if it did occur, denial even of the importance of survivability. At its most optimistic this spirit of nostalgia has produced the notion that, even against very advanced threats, some form of basing in silos or shelters on the earth's surface can be made survivable for the long haul. This has spawned ingenious designs of great intricacy -- the strategic equivalent of Swiss clocks (or, for the unkind, Rube Goldberg machines): multiple shelters and shell games, superhard silos, reliance on fratricide effects, layers of boulders to defeat attacking earth penetrator warheads, and so on.

If one persists in silo or shelter basing for the long term, however, and begins to consider the difficulty of dealing with long-run threats against such known fixed points on the surface of the earth, one is driven inexorably to consider augmenting these Swiss clock basing modes with some form of ballistic missile defense. It is one of the great ironies of the strategic debate over

the last decade that the Carter Administration
favored the form of ICBM basing (multiple shelters
with deception) that most lent itself to ballistic missile defense, but did not favor such
defense. The Reagan Administration, on the other
hand, while supporting new departures in ballistic
missile defense, has gone to great lengths, twice,
to avoid the mode of basing ICBMs that would have
made ballistic missile defense most practical. A
combination of the long-run threat to fixed points
on land and the operation of our political system
thus seems to be leading us away, in the long
term, from silo or shelter basing for our ICBM
force.

Whether deep underground, in hardened mobile
launchers, in specially designed aircraft or in
some as-yet-unimagined basing mode, as we approach
the turn of the century there will most likely be
a very different sort of ICBM force than Titans,
Minutemen, or MXs, sitting in their silos. The
chapters that follow well describe the difficulties, complexities, agonies, and uncertainties in
this fascinating evolution of a vital strategic
system.

Acknowledgments

We would like to express our deep appreciation to Amy Bodnar, Jill Coleman, Candace Dickerson, Jeanine Ellars, Beth Miller, and Sue Munn for their long efforts in preparing this manuscript. We also would like to offer a special thanks to Josephus Briggs of the National Institute for his helpful comments on much of the manuscript.

The Editors

1
The New Missiles: Putting the Debate into Proper Perspective

Keith B. Payne and Barry R. Schneider

The debate over U.S. ICBM modernization has been long and often confused. During the last decade endless basing modes for the Peacekeeper (MX) missile have been examined. Decisions in the Reagan Administration and public opposition in Utah and Nevada scrapped the multiple protective shelters (MPS) basing mode for Peacekeeper ICBMs favored by the Carter Administration. The U.S. Congress has been reluctant to fund subsequent Peacekeeper basing modes such as the closely-spaced basing system proposed two years ago. Throughout the debate, the cast of political and military decision-makers have changed. Four Presidents have sat in the White House since the Peacekeeper program was launched. Half a dozen Secretaries of Defense have come and gone. Several sets of Joint Chiefs of Staff have debated and redebated each successive ICBM basing mode, program change, and strategy change. More than half the members of Congress who originally voted R&D start-up money for the Peacekeeper have since departed Capitol Hill. Each new Congress debates the issues surrounding ICBM modernization in a manner that suggests that ICBM modernization issues are unfamiliar to most members.

Too often, in the Congressional as well as the public debate, the wrong questions have been discussed or inadequate responses have been given to the correct questions. All too often, inordinate weight appears to be attached to narrow technical issues while too little has been assigned to the key role of the ICBM in U.S. deterrence and arms control policy.

This book identifies some of the fundamental questions that should be at the heart of any meaningful debate over what direction U.S. ICBM modernization should take. It addresses, for example, how and whether or not the Peacekeeper ICBM should be deployed, and -- more fundamentally -- whether the U.S. should maintain an ICBM force at all. It examines whether a dyad of strategic bombers and submarine launched

1

ballistic missiles would be any less effective for the maintenance of stability than the traditional triad of strategic forces which includes ICBMs.

A key element in this book, and an argument largely missing from the U.S. public debate, is how the United States seeks to deter the Soviet Union from attacks on the United States, its allies, and vital western interests, and what kind of U.S. strategic force is most likely to deter war. Most importantly, these essays examine the role that large ICBMs and small mobile ICBMs might play in keeping the peace between the Soviet Union and the United States. Such fundamental questions are raised as, "what targets does the U.S. need to be able to threaten to deter Soviet military action?" "What importance do new U.S. ICBMs have for supporting the current, official policy of deterrence," and "does this deterrence posture make sense?"

Questions such as these go to the heart of the strategic debate in the United States. Unfortunately, within the current debate about ICBM modernization, too little of the discussion pertaining to Peacekeeper and small ICBMs (SICBMs) has been placed in the comprehensive framework of these questions. Individual issues, although indeed critical, have been viewed in isolation from this broader framework -- a tendency which has produced more misunderstanding than clarity.

For example, the issue of "survivability" has rightly been a primary concern of Peacekeeper development. A goal of ICBM modernization is to improve U.S. ICBM survivability. Yet two basing modes which would have provided increased survivability, multiple protective shelters (MPS) and closely spaced basing (CSB) of superhardened silos did not receive adequate support to be sustained. The subsequent decision to place 100 Peacekeeper ICBMs in existing silos was then derided as inconsistent with the goal of ICBM modernization, i.e., enhanced survivability. Unfortunately, focusing on a single issue, even one as important as ICBM survivability, distorts the context within which the debate over ICBM modernization should take place and obscures the fact that survivability has never been the sole rationale for ICBM modernization. Another U.S. objective was to achieve essential equivalence in strategic power by offsetting the Soviet advantage in heavy counterforce potential. Even more fundamental is the U.S. need for forces that can credibly threaten the things that Soviet leaders value most in order to deter them from attacks on the U.S., its allies, and its vital interests. Especially important in this regard is the capability to threaten Soviet military and political facilities which have been extensively hardened against nuclear attack.

The reason that Peacekeeper based in existing silos would provide much needed support for the current U.S. deterrence strategy has received far too little attention. Many members of Congress (and the public) rightly have questioned why the deployment of Peacekeeper is necessary in the absence of plans to provide increased ICBM survivability. The answer has always been clear, although rarely provided. Peacekeeper deployment will permit the U.S. to hold at risk the political and military control facilities of the Soviet state and CPSU that represent the highest values of the Soviet leadership. The official strategy to support deterrence reflects the belief that these highest value assets should be held at risk to deter Soviet attack on the U.S. and its allies.

The fact that the Soviet Union has deployed over 700 fourth generation ICBMs since the mid-1970s is not reason enough to support Peacekeeper deployment. The U.S. should not, and does not, mimic the Soviet Union if that government chooses to deploy weapons in an ill-advised fashion. The Soviet ICBM force represents a very ominous threat to the U.S. However, we plan to deploy far fewer counterforce capable ICBM weapons than has the Soviet Union, and U.S. deployment plans are designed to satisfy the specific deterrent requirement of posing an adequate threat to those hardened military and political control facilities that constitute the highest values of the Soviet leadership.

It is necessary to properly assess how deterrence policy, targeting requirements, and arms control con-siderations interact in order to acquire a coherent perspective on the issues surrounding ICBM moderniza-tion. Focusing on any single consideration while neglecting others and their overall strategic context will continue to obfuscate a matter of utmost signifi-cance for national security.

Fortunately, the bipartisan Scowcroft Commission on Strategic Forces, provided an invaluable service by identifying the key issues related to how and why we should modernize U.S. ICBMs and by examining these issues within a comprehensive framework. The major recommendations of the Commission's report, that Peace-keeper be deployed in silos and that this deployment be complemented by subsequent development of a small mobile ICBM, were derived from a comprehensive view of policy, strategy and force requirements. Yet that report, issued in April 1983, could not anticipate all of the subsequent political and technological develop-ments. The debate over ICBM modernization now appears to be in jeopardy of returning to a narrowed focus on discrete issues which should not be addressed in isola-tion from the broad strategic framework identified by the Scowcroft Commission.

This book represents an effort by analysts at the National Institute for Public Policy to provide a comprehensive examination of U.S. ICBM modernization issues. The individual chapters of this book examine important pieces of the broad policy and strategic landscape that is critical to an understanding of ICBM modernization and the role of the ICBM in U.S. deterrence policy. It is our hope that this effort will help to clarify how important individual issues such as arms control and missile survivability relate coherently to ICBM modernization, and how U.S. ICBM modernization will, in turn, stabilize the strategic environment.

The second chapter of this collection of essays begins with an examination of the reasons for a diverse mix of U.S. strategic forces and provides commentary on the conclusions of the Report of the President's Commission on Strategic Forces. The virtues of multiple basing modes within the ICBM force are discussed, as well as the utility of maintaining a triad of U.S. strategic retaliatory force employing land, sea and air launched systems. Examined are the reasons for integrating U.S. ICBM modernization plans with U.S. arms control initiatives, and the part to be played in deterrence by a tandem of silo-based, multi-warhead Peacekeeper ICBMs and by single-warhead, small, mobile ICBMs.

The third chapter looks at the potential of hard mobile launchers (HMLs) and deep underground basing for fielding a truly survivable and thus more effective U.S. ICBM deterrent force. Development of hard mobile launchers is an important requirement for the successful deployment of a mobile small ICBM force on U.S. military reservations. This report focuses on the issues raised by the HMLs and the questions that need answers concerning SICBM. Deep basing of ICBMs is also studied as a possible future basing mode for U.S. ICBMs. The novelty of this basing mode is perhaps its greatest political handicap. Although deep basing has the greatest potential for fielding virtually invulnerable ICBM forces in the future, the concept does not readily lend itself to the promptness of response now considered a major asset of ICBMs. The essay in Chapter 3 also discusses the potential of deep basing for protecting not only ICBMs but also the U.S. National Command Authority, its associated C^3I instruments, and other strategic force assets.

The fourth chapter examines the strategic implications of the introduction of silos superhardened to potentially 25 times the blast resistance of current U.S. silos. Discussed are the potential changes that superhardening will require in the mix of U.S. strategic offensive forces, in U.S. plans for active

ballistic missile defenses, in U.S. deterrence and
targeting policy, in U.S. arms control policy, in
Soviet offensive countermeasures, and in Soviet ICBM
deployment choices. Superhard structures may be one
effective answer to very accurate counterforce weapons,
and the superhardening breakthrough may help reverse
the past trend toward increased vulnerability of fixed
targets to missile strikes.

The fifth chapter examines the likely implications
of President Reagan's Strategic Defense Initiative and
of active strategic defenses for future U.S. ICBM
programs. The analysis focuses on how future effective
ballistic missile defenses might change the terms of
deterrence strategy from an emphasis on punitive
retaliatory threats against Soviet assets to an
emphasis on defenses capable of denying wartime victory
to the Soviet Union -- even if U.S. retaliatory capabilities were also drastically reduced by Soviet
strategic defenses. Finally, this chapter discusses
the continuing requirement for modern ICBM forces
during any defensive transition.

The sixth chapter defines what, in theory, an
ideal ICBM force ought to be able to contribute to
stability, to military capability, and to arms control
progress. It then evaluates two options: (1) deploying only small ICBMs or (2) deploying both Peacekeeper
and small ICBMs. The analysis shows that the deployment of two types of new ICBMs comes much closer to
providing the advantages of an ideal ICBM force than
the deployment of only small ICBMs, without the targeting footprint of Peacekeeper ICBMs.

The seventh chapter examines the uncertainties
confronting any Soviet warplanner given the task of
evaluating, in advance, the outcome of any Soviet first
strike against U.S. silo-based ICBMs. Identified and
discussed are Soviet uncertainties about the likely
U.S. strategy when provided with tactical or strategic
warning of the Soviet attack. Soviet uncertainties
about missile performance, attack coordination, fratricide, silo hardness, meteorology and other target-end
factors could make forecasting of attack outcomes very
difficult indeed. Uncertainty, for the most part,
appears to be a major ally of war deterrence. This
chapter shows how silo-based ICBMs like Peacekeeper and
Minuteman III can contribute significantly to stability
despite the "window of vulnerability" predictions of
the past.

The eighth chapter of this book is devoted to how
Soviet officials appear to perceive the U.S. ICBM
modernization programs and U.S. strategic motives.
Examined is the relative impact of programs such as the
Peacekeeper or small ICBMs on Soviet military programs
and doctrine, and on Soviet arms control and propaganda

responses. The difference between U.S. and Soviet perceptions of the rival strategic programs, rival strategies, and views on how nuclear conflicts might unfold are highlighted. Chapter nine provides some concluding comments on the role of the Peacekeeper in the maintenance of stability.

This volume was written by analysts at the National Institute for Public Policy as a contribution to the present very important debate over how the United States should proceed into the late 1980s and 1990s with new intercontinental ballistic missiles. If it adds to the understanding of the public and their elected representatives, it will have served its purpose.

2
The Virtues of Diversity and the Future of the ICBM

Colin S. Gray and Blair Stewart

There are at least four questions concerning ICBM modernization that any serious debate on the subject should encompass:

(1) What is the need for ICBMs and for a strategic triad?

(2) What is the rationale for putting Peacekeeper ICBMs in silos?

(3) Is there a need for small ICBMs to supplement U.S. silo-based ICBMs?

(4) How can ICBM modernization be most effectively combined with arms control to produce a coherent policy and a national consensus?

The first question will be dealt with at length here because it is important that decision-makers and the public understand the virtues of diverse U.S. strategic forces. The other three questions have been dealt with at length by the Scowcroft Commission and their report to the President deserves, and receives here, an evaluation as to its strategic and political merit.

Why ICBMs? Why a Triad? The Virtues of Diversity

Some people seem to believe that simply because an idea has been in circulation for many years, it must have outlived its usefulness. There is no inherent virtue in the novelty of an idea. It is unfortunate, though perhaps culturally inevitable, that the U.S. domestic debate on nuclear strategy repeatedly produces ill-considered challenges to the triadic structure of U.S. strategic forces.

Portions of this chapter first appeared in an article by Colin S. Gray, "Abiding Realities and Strategic Needs," Air Force Magazine, vol. 66, no. 7, July 1983, pp. 73-77.

From time to time entire classes of military technology, and even whole arms of a military service, are
overtaken by new technologies and tactics. However,
the triadic structure of strategic forces has yet to
join their number.

Prominent among the virtues of a diversified strategic forces' posture is the very obvious fact that it
has done its job successfully to date. A more simply
structured force posture might perform as well, but --
given what is at stake -- why should the risk be taken?
It is more likely than not that the elimination of one
leg of the triad, the ICBMs for example, would result
in the U.S. Government seeking compensation in the form
of a more muscular dyad of strategic submarines and
bomber forces. Dollar cost savings likely would be
nonexistent or trivial.

To the very limited extent to which a strategic
triad of forces provides redundant capabilities, an
analogy with elevator safety design features is appropriate. An elevator accident could be so catastrophic
for those involved that backup systems to backup
systems are provided for safety. No elevator designer
is permitted to ask of safety engineering, "how little
is enough?" Statistically improbable sequences of
events do occur. The designer of strategic forces
knows that the potential failure of one element of the
triad needs to be insured against by the existence of
complementary retaliatory forces in the other triad
legs.

Survivability Through Diversity

A diverse force structure enhances the survivability of U.S. strategic forces. The dispersal of the
U.S. strategic nuclear arsenal among ICBMs, SLBMs and
aircraft poses extremely severe, and perhaps impossible, difficulties of attack timing for a first-strike
planner. Even if they are in a peacetime posture of
low alert, half of the U.S. SSBN fleet will be at sea,
about 30% of the SAC bomber force will be on runway
alert, and the ICBM force -- as always -- will be
instantly ready to fire.

Each element of the triad has distinctly different
characteristics which require the Soviets to use unique
methods to negate each. For example, U.S. ICBMs at
present are not vulnerable to Soviet SLBM weapons,
since the hardness levels of their silos is sufficient
to withstand the accuracy and yield combinations of
Soviet SLBMs. Accordingly, the Soviet Union must plan
to attack ICBMs directly with either their own ICBMs or
bombers which must penetrate U.S. airspace or launch
cruise missiles from a stand-off position.

SLBM-carrying submarines are virtually invulnerable due to their vast ocean patrol areas and the

current state of ASW technology. This compels the
Soviet Union to adopt a strategy of "one-on-one"
tracking if they wish to negate U.S. missile submarines. This, in turn, makes a coordinated, successful attack on the U.S. SSBN force an unlikely scenario
-- not only would all Soviet attack submarines need to
be in place to initiate the attack, but the Soviet
Union would also have to worry about the difficult
problem of command and control linkage to coordinate
the attack.

Finally, U.S. bombers, particularly those that are
on runway alert, have the ability in most cases to be
launched subsequent to detection of a Soviet SLBM or
ICBM launch and to be airborne prior to arrival of
ballistic missile weapons. This attribute forces the
Soviet Union to: (1) deploy their SLBM submarines
relatively close to the continental United States if
they wish to barrage potential bomber escape paths with
enough overpressure to kill the aircraft; (2) rely on
prepositioned special forces prepared to act when they
are called upon to do so; or (3) be prepared to defend
against those bombers which escape when they penetrate
Soviet airspace. Given the massive Soviet investments
in air defenses, it is obvious the Soviets consider the
last option the most realistic.

Not only do these individual characteristics
necessitate unique attacks to counter a single triad
element, they also provide complementary survivability
for the triad as a whole.

If the Soviet Union wishes to launch a coordinated, effective ballistic missile attack on U.S.
bombers, ICBMs, SLBMs in port, and U.S. strategic
command and control, it must either launch its ICBMs
and SLBMs simultaneously or stagger their launches so
that the SLBM and ICBM weapons arrive at the same time.
This problem is caused by the different times of flight
associated with the two systems. Soviet ICBMs, based
in the interior of the Soviet Union, require about
25-35 minutes to reach targets in the central United
States. On the other hand, Soviet SLBMs, whose submarines must launch their missiles from home bases on
the northern periphery of the Soviet Union, or from
their patrol areas, require about 15-25 minutes to
reach U.S. strategic targets.[1] This results in a
difference in Soviet ICBM and SLBM flight times of ten
to twenty minutes.

Given this differential, the Soviet Union faces a
monumental attack coordination problem. If the Soviets
elect to launch their SLBMs and ICBMs simultaneously,
they must recognize that their SLBM weapons will arrive
in advance of their ICBMs weapons. Thus, there would
be a period of at least ten minutes between the initial
detonations of Soviet SLBM weapons on U.S. territory

and the detonations of Soviet ICBMs on their intended targets. On the other hand, if the Soviets elect simultaneous detonations on targets in the U.S., they must delay the launch of their SLBMs by ten to twenty minutes in order to coordinate their arrival times with those of Soviet ICBMs.

In the first case, Soviet SLBMs would already have detonated on continental U.S. bomber bases and SLBM submarine ports before Soviet ICBM weapons arrived at U.S. ICBM silos. Thus, the U.S. National Command Authority (NCA) would have positive proof of a Soviet nuclear attack on the United States, and could act to launch ICBMs before they could be destroyed in their silos.

In the second case, no Soviet weapons would arrive on their U.S. targets until 25-35 minutes after the first launch. U.S. bombers on runway alert, therefore, would have ample time to take off and be well on their way to targets in the Soviet Union before their bases or escape corridors could be attacked.

In this light, then, a highly successful attack on the entire U.S. triad would be extremely difficult to achieve and, more importantly, would entail a tremendous amount of risk for the Soviet Union. Combining the synergistic survival traits of the bomber and ICBM legs of the triad with the fact that a healthy portion of the SLBM forces are always at sea, the Soviet Union must always face the prospect that no matter how it chooses to attack U.S. strategic forces, there is little margin for error and a high probability that the United States would be able to respond with devastating results.[2]

To achieve their offensive and defensive goals, Soviet force planners are compelled to seek to counter each leg of the triad. It is more difficult to counter ICBMs, submarines, and air-breathing vehicles, than it would be if the structure of U.S. forces were reduced to a dyad or monad. By commanding the attention of a noteworthy fraction of scarce Soviet defense assets, each leg of the triad contributes to the survivability of the other two legs.

A diversified U.S. strategic force structure places additional burdens on a Soviet strategic force which also must worry about regional foes other than the United States. These potential problems force the Soviets to always be concerned that in any attack on the United States they are able to hold in reserve sufficient strategic weapons to provide an adequate backup to their intermediate and tactical nuclear forces. The large number of weapons required for a successful attack on the U.S. diminishes the number of weapons the Soviet Union can hold in reserve for other adversaries and for unforeseen contingencies. Conversely, the requirement to hold portions of their strategic force in reserve reduces the number of weapons the Soviet Union can target on the U.S. strategic force, thereby improving the prospects that portions of that force would survive a first strike.

Diversity enhances the survivability of U.S. strategic forces by allowing the United States to hedge against an unforeseen Soviet technological breakthrough or a system-unique failure which could affect an entire element of the strategic force. For example, if the United States chose to deploy only missile-carrying submarines, it would run the risk of the Soviet Union being able to concentrate its resources on anti-submarine warfare (ASW) without having to pursue the additional systems and capabilities that would be required in order to defeat other types of strategic forces.

The difficulties in locating and tracking submarines could yield to scientific advance on short notice. That is not a prediction, but neither is it a wildly fanciful notion. A United States with a very muscular-seeming dyad of Poseidon and Trident ballistic missile submarines and manned bomber and cruise missile carriers could discover belatedly, that it did not have a second-strike deterrent at all. SSBNs away from port may be relatively invulnerable, but they may not remain so forever.

The truth is that nobody knows. The prudent U.S. defense planner in the mid-1980s knows that the Soviet Union confronts no fundamental problems of science in countering air-breathing vehicles, and he must assume that the odds against SSBNs at sea enjoying a true permanence of invulnerability are rather long. Experience suggests that the long lead-time necessary to weaponize new technical developments and absorb them into Soviet forces in useful numbers, should yield all of the notice the U.S. would need to take countermeasures. But, the prudent U.S. defense planner who has read his military history will know that genuine tactical, rather than technical surprise -- on such a scale that it has strategic results -- has happened many times in the past and could happen again.

Technical or tactical surprise cannot be dismissed, but a strategic triad minimizes the danger that any unfavorable surprises would have great strategic significance. Additionally, if a problem developed in the subsystems of either the submarine or the fleet ballistic missiles themselves, the United States would run the risk of having its entire deterrent force degraded until the problem could be corrected.

U.S. strategic force diversity provides a security net for such eventualities. First, when an individual triad element becomes vulnerable to Soviet capabilities (such as is perceived to be the present case with U.S. ICBMs) this condition does not constitute triad vulnerability. Second, when technical problems have arisen in one triad element (such as they have with certain subsystems of SLBMs), the forces comprising the remaining two elements have been able to maintain the number of weapons on alert which could attain U.S. targeting objectives if conditions so warranted.[3]

For the future, the planned strategic force modernization program should further enhance the survivability of U.S. strategic forces. The B-1B will decrease the escape time required by U.S. bombers, thereby making the timing problem for a coordinated Soviet attack on U.S. bombers and ICBMs even more difficult. The B-1B also will insure that the U.S. bomber force can continue to achieve high penetrability against Soviet air defenses. As B-52 aircraft are phased out, and if the "Stealth" technology does not prove operationally feasible and cost effective, the effectiveness of the B-1B as a penetrating bomber could be further extended.[4]

The small ICBM will enhance triad survivability by exacting a fairly high attack price for its neutralization. This in turn will further strain Soviet weapon allocation and the problems facing Soviet targeteers. Additionally, the improved in-flight hardness of both the Peacekeeper and small ICBMs to nuclear radiation, dust, and electromagnetic pulse will make these missiles more survivable against any potential Soviet combination of endo- or exo-atmospheric "pin" attack and simultaneous attack against ICBM launch sites.

Finally, the increased range of the Trident D-5 SLBM will extend the patrol areas of U.S. Trident submarines, thereby making them even more survivable against Soviet ASW efforts.

In summary, diversity enhances the survivability of U.S. strategic forces by requiring the Soviet Union to employ unique methods and strategies for a successful attack on any one element of the U.S. strategic force. These individual characteristics and the problems they create for the Soviet Union act synergistically to make a coordinated, effective attack on the U.S. triad an extremely difficult and risky undertaking. Hence for the predictable future, the triad concept increases the probability that a significant portion of U.S. strategic forces will survive a Soviet attack.

Greater Effectiveness Through Diversity

The preceding discussion outlined the defensive synergism provided by the diversity of the triad: those that increase the prospects that sufficient

U.S. strategic forces would survive a Soviet first strike to assure an unacceptable U.S. retaliation. But diversity also enhances the offensive effectiveness of U.S. forces. For example, U.S. SLBMs and ICBMs could neutralize Soviet air defense interceptors opposing U.S. bombers by attacking their bases and control network. The interceptors might survive, but they would be out of fuel and without radar control prior to the arrival of U.S. bombers. At the same time, U.S. SLBMs and ICBMs have the capability to disrupt and severely damage the fire control network for Soviet SAM launchers.

When viewed in the aggregate, therefore, the U.S. triad presents to the Soviet Union a robust, effective force structure.

The maintenance of a triad means that U.S. weapon designers do not need to provide a single all-purpose superweapon. There is a wide range of target categories in the U.S. target inventory, some of which are far easier to destroy or damage than are others. The simpler the strategic force structure, the greater the versatility that would need to be built into each weapon system. Cross-targeting against high value targets among delivery vehicles from different legs of the triad, provides high assurance of achieving requisite damage expectancies and enhanced target coverage. It guards against the failure or vulnerability of any single leg of the triad. As the Scowcroft Commission observed, U.S. strategic policy has been plagued in recent years by unreasonable demands that a single weapon be able to solve a set of complex problems that, in the strategic context of the 1980s, do not lend themselves to a single solution.

A strategic forces triad enables the United States to develop the flexible arsenal suited for: (1) prompt counterforce duty; (2) delayed counterforce missions; and (3) the posing of a large strategic reserve threat against the entire target structure of an enemy as a tactic for escalation control.

ICBM Characteristics and Capabilities

ICBMs, which comprise over half of the U.S. day-to-day alert weapons and more than two thirds of the alert nuclear delivery vehicles, are home-based and thus effectively secure from conventional attack during non-nuclear war.[5] This home basing also raises the threshold of nuclear war, since to attack U.S. ICBMs the Soviets must also attack the continental United States.

Soviet military writings and Soviet investments in ICBMs strongly indicate the Soviets regard ICBMs as the dominant strategic system. The Strategic Rocket Forces (SRF) are the elite military service in the Soviet armed forces. Accordingly, the Soviet Union perceives an ICBM-armed United States as a more powerful adversary which possesses a significant number of the very weapons they themselves would rely on most heavily in the event of a war.

In addition, ICBMs have certain unique operational capabilities which contribute directly to Soviet perceptions of U.S. strategic force effectiveness. The ability of ICBMs to be launched promptly against command and control systems and military forces given the United States a capability to interrupt the execution of a Soviet offensive operation. With their rapid retargeting and selective employment capabilities, ICBMs insure that the United States can concentrate force against Soviet offensive operations when and where it is most needed. In effect, this capability offers to the Soviets the prospect that the U.S. would be able to defeat their strategy for nuclear, or even conventional, war.

For the predictable future, ICBMs will possess the best prompt countermilitary capability of all U.S. strategic forces: their combination of weapon yield, accuracy, quick reaction time, and short flight time gives the United States the ability to place at risk most of the assets which are important to the conduct of Soviet war operations -- including ICBMs which the Soviets will have withheld from an initial attack. The ICBM's ability to attack promptly and destroy key elements of Soviet offensive forces also provides the greatest measure of damage limiting counterforce capability to the United States, and the greatest prospects for early war termination.

The communications links to U.S. ICBMs are very secure and redundant. They insure that ICBMs have the highest probability of command and control connectivity during wartime. Not only are these links more secure against nuclear effects, they are more immune to spoofing, interdiction, and interrupting by Soviet agents or jamming systems.

Due to their high reliability and low manning requirements, ICBMs maintain the highest alert and readiness rates of all the triad elements. The high alert rates and quick responsiveness of U.S. ICBMs combine to further discourage a surprise first strike on the United States by forcing the Soviet Union to worry about vulnerability of their own strategic forces -- a good portion of which are not on day-to-day alert -- to prompt U.S. retaliation.

To illustrate, if the Soviets decided to attack the United States, they would want to have available to them as much of their strategic force as possible to insure a high confidence, high success attack. But if they generate their forces, the Soviets may alert the

United States, sacrifice the element of surprise and
diminish the effectiveness of a surprise attack. Thus,
contrary to a widely held belief, highly responsive
U.S. ICBMs inhibit rather than encourage Soviet first
strike planning.

ICBMs also offer cost advantages not available in
the other triad elements. For example, ICBMs today
represent only about 12% of the triad operations and
support costs. Savings which are accrued in the operation of ICBMs can thus be applied toward maintaining
higher alert rates in the other triad elements, thereby
increasing their deterrent effectiveness.

Finally, ICBMs along with SLBMs have the greatest
probability of penetrating to their targets and are the
most difficult strategic weapons against which to
defend. Their hypersonic reentry velocities and low
radar cross sections, coupled with possible penetration
aids such as chaff clouds and decoys, make terminal
defense against these weapons most difficult.

When it is deployed in the latter part of this
decade, the Peacekeeper missile will greatly improve
the effectiveness of the U.S. ICBM force. First,
because it represents a new generation ICBM, Peacekeeper will provide a hedge against unforeseen failures
in its present aging Minuteman force. Second, Peacekeeper's increased range (almost 1,000 nautical miles
more than Minuteman III) and increased footprint (about
twice the cross and downrange capability of Minuteman
III) give it greater operational and targeting flexibility. This will allow the U.S. ICBM force to hold at
risk the high priority, proliferated structure of
Soviet hardened military and leadership assets, particularly those in the southern-most regions of the
U.S.S.R. Third, its rapid retargeting potential in a
crisis combined with its effectiveness against all
target types, should render Peacekeeper invaluable in
achieving U.S. nuclear weapon employment objectives in
the event of a nuclear war. Fourth, the improved
throwweight of the Peacekeeper will provide future
flexibility in the use of penetration aids, maneuvering
reentry vehicles, or higher yield warheads if these
became necessary to preserve the effectiveness of the
U.S. ICBM force.

Bomber Characteristics and Capabilities

The manned bomber also possesses unique characteristics which enhance the perceived effectiveness of
the U.S. triad. Bombers are the only triad systems
which can be recalled once they have been launched
against their targets. This provides a hedge against
accidental weapons commitment in the event the NCA had
incomplete information regarding a nuclear attack. In
addition, this feature allows the NCA to delegate the

decision to launch the bomber force to the Commander in Chief of SAC (CINCSAC). This enhances their survivability by allowing the bombers to be launched at the attack warning level CINCSAC believes necessary for their survival, or if he is temporarily cut off from communication with the NCA.

Up to a certain point in their flight profiles, bombers can also be retargeted if warranted by an initial assessment of the course of events in a nuclear war. If they no longer can communicate with higher military authorities, aircraft commanders have the ability to perform a limited amount of battle damage assessment and, if their primary target has already been destroyed, proceed to secondary targets.

Finally, the sophistication of the U.S. bomber force compels large Soviet expenditures for air defenses: expenditures which otherwise might be diverted to more threatening, offensive weapons. The fact that U.S. bombers can attack the Soviet Union from a number of different directions requires the Soviets to cover all possible bomber attack routes with a comprehensive, in-depth air defense system. The operation and maintenance of this system also consumes large amounts of manpower and rubles which could be used elsewhere were it not for the U.S. bomber threat.

The B-1B will increase significantly the effectiveness of the U.S. bomber force through its improved ability to penetrate Soviet air defenses. This improved effectiveness is a function of the B-1B's lower radar cross section, improved avionics and electronic countermeasures capabilities, improved nuclear hardness, improved aerodynamic characteristics, and improved weapons capacity. Additionally, the B-1B will enter the U.S. inventory at a time when the average B-52 force may be approaching the point of "marginal return" in terms of the U.S. ability to maintain a significant portion of the aircraft on day-to-day alert. If unforeseen B-52 maintenance or reliability problems should arise, the fact that the B-1B is already in production should provide the United States with a good hedge against the reduced effectiveness of a significant portion of its bomber force.

SLBM Characteristics and Capabilities

The nuclear submarine platforms for SLBMs provide these weapons with the highest survival probability of the current triad elements. Given the extreme difficulties associated with conducting open-ocean search operations for submarines, this high survivability should continue for a long time.[6] This survivability makes SLBM weapons ideally suited for constituting the bulk of a U.S. secure reserve force which would not be launched in an initial U.S. retaliatory attack. With

the addition of the extended range D-5 missile to the
Trident submarines, the patrol areas for SLBM submarines could be expanded, thereby making them even
more survivable against Soviet ASW efforts.   Although
U.S. SLBMs currently do not possess the necessary
accuracy to threaten hard targets like ICBM silos, the
D-5 missile and Trident submarine system should provide
the United States with sufficient capability at sea to
threaten all but superhardened Soviet targets.   This
will provide a hedge against future extension of
protective hardening by the Soviets.

The submarine offers a mobile platform from which
to launch SLBMs on a number of different attack
azimuths against the Soviet Union, thereby complicating
the potential ability of the Soviet Union to mount an
effective ballistic missile defense (BMD) system.   The
addition of the D-5 missile to Trident submarines will
increase their flexibility and further complicate the
Soviet BMD problem.

Merits of a Diverse Force Structure

In summary, diversity enhances both the defensive
and offensive synergism of the U.S. strategic force.
For over twenty years, this diverse force structure
has been perceived by the Soviet Union and the world as
a powerful and effective deterrent force.   If the
United States were to abandon one or more elements of
the triad, it is obvious that the Soviet Union would be
better able to counter the remaining force structure
than it can now.   Additionally, it is doubtful that the
world would perceive a dyad or monad, no matter how
strong either was, as the equivalent in deterrent
effectiveness of the U.S. triad.

The distinctive qualities of the ICBM, as con-
trasted with other classes of strategic weapons, are
the qualities most appropriate to hold at very
plausible risk the highest value assets of the Soviet
state.   Put in cross-cultural terms, the U.S. protects
its highest value (people) by holding at risk the
highest values of the Soviet Union (the political
control structure and major elements of its most potent
military capability).   Cruise missiles, penetrating
bombers, and SLBMs all have valuable synergistic roles
to play for deterrence, but they cannot today, or
prospectively tomorrow, provide a level of dissuasion
capable of substituting for the ICBM.   Abandoning the
ICBM component of the triad, and moving to a less
diverse U.S strategic force could only lead to an
increased probability that U.S. foreign policy would
have to be conducted in the shadow of Soviet strategic
power which was perceived by the world to be superior
to that of the United States.

Six points about the triad merit emphasis. First, each leg of the triad has distinctive strengths and weaknesses. The separate legs are not competitors for the mission of strategic nuclear deterrence. Second, the "performance" of the entire triad, in peacetime, crisis-time, or war, is greater than the sum of its parts (to remove the ICBM or the bomber/CMC leg would be to effect a disproportionate weakening in overall strategic capability). Third, much of the overlap in capability among the triad legs is desirable and is not evidence of a wasteful redundancy. For example, U.S. policy requires our forces to be able to pose a credible threat to very hard Soviet targets both at the outset of a conflict and throughout a conflict, which is why the U.S. plans to deploy both Peacekeeper and Trident D-5.

Fourth, there is a lot to recommend the folk wisdom which holds that "if it ain't broke, don't fix it." There is a military and technical logic to the triad, recognized in Soviet as well as U.S. practice, that far transcends bureaucratic or policy inertia as an explanation of why the strategic force posture has been consistent for more than two decades. The triad is familiar, it is responsive to U.S. policy needs, and there is an absence of persuasive reasons for altering it. In short, it works.

Fifth, as noted already, the logic of the different capabilities embraced within the triad is in accord with U.S. policy guidance. Critics of particular ICBMs or of particular manned bombers need to be clear in their thinking as to whether (1) they are challenging an individual weapon system or an entire class of weapon systems, and whether (2) their discontent should be focussed upon weapons or upon the policy that the weapons are designed to express and execute. U.S. policy guidance has been well ahead of technical achievement, let alone deployed forces, for more than a decade.

Finally, in both a political and technical sense, a triad structure for strategic forces is healthy for strategic stability.

From this analysis it should be clear why ICBMs are important to U.S. security and why a three-legged strategic structure is more stable than a two-legged one. Now let us turn to the questions as to whether silo-basing of ICBMs makes sense, if so whether silo-based ICBMs need to be supplemented with small mobile ICBMs, and what is needed to bring about a bipartisan consensus behind a U.S. strategic program that coherently integrates ICBM modernization with U.S. approaches to arms control. For this examination the recent Scowcroft Commission Report is a useful point of departure.

The Scowcroft Commission and the Future of the ICBM Program

In the conclusions to their recent report, the President's Commission on Strategic Forces encapsulated the major problem in the following words: "Finally, the Commission is particularly mindful of the impor- tance of achieving a greater degree of national con- sensus with respect to our strategic deployments and arms control."[7]

Although it is too early to claim definitive success, at least there is a strong probability, on current evidence, that the Commission may have achieved what seemed almost impossible only a short time before -- the forging of a politically workable consensus that could endure and is generically supportive of the ICBM program. Without playing down the contentiousness of some of the details, the carefully crafted package of arguments contained in the report have effected what appears to be a sea-change in the terms of debate over ICBMs. Early in 1983 the public debate was shifting rapidly from the issue of "which Peacekeeper basing mode to choose," to the fundamental question of "why does the United States need an ICBM force at all?"

As critics of the Reagan Administration have insisted that the President take proper account of all of the recommendations of his Commission -- especially with particular reference to its arms control arguments -- so, in its turn, the Administration is able to insist that the Congress, too, take all the recommenda- tions into due account, especially the report's strong endorsement of Peacekeeper deployment. Prominent among the many virtues of the Commission Report is its skillful interweaving of elements that are very dear to the hearts of different constituencies. However, given the heat that ICBM and arms control controversies have generated over the past several years, and given the skepticism of others' motives that pervades the ranks of those who debate these issues, the interwoven character of the arguments in the Report is a potential source of major weakness, as well as an element of strength. Everyone prefers a Chinese menu, from which they can pick and choose those items that they prefer. The Commission Report is a set meal with no substitutes permitted as yet. This situation may change, of course, if the Soviet Union ever returns to the negotiating table and responds to the Commission- inspired START proposal, and as critically important details concerning the new, proposed small ICBM begin to emerge.

Whatever one thinks of the merits of the Commission Report, if nothing else it would seem to have created a more bipartisan consensus backing ICBM modernization and related issues of arms control. By any standard, that is a major achievement -- given the terms of the debate following Congressional rejection of closely spaced basing for the Peacekeeper in December 1982.

The Commission Report: Changing the Question

The Commission decided, early on, that the question that had been central to the ICBM debate for the previous four years could not be answered in a positive way. Previously debated had been this question:

> How can a force consisting of relatively large, accurate land-based ICBMs be deployed quickly and be made survivable, even when it is viewed in isolation from the rest of our strategic forces, in the face of increasingly accurate threatened attacks by large numbers of warheads -- and how can this be done under arms control agreements that limit or reduce launcher numbers?[8]

The Scowcroft Commission did not so much pass negative judgments on the array of Peacekeeper basing modes that have been the focus of recent controversy, as rather change, in fact broaden, the question. The question was reformulated so as to ask what kind of a total mix of strategic forces should the United States sustain and develop in order to deter a uniquely Soviet type of adversary. The case for ICBM's in general, and the Peacekeeper in particular, flowed naturally and very persuasively from that analysis.

Prior to its discussion of the details of the ICBM program, the Commission Report provides a compelling and relentless argument regarding the requirements of deterrence. The Report broadens the sense in which strategic stability typically is discussed, away from examining the possible implications of theoretically vulnerable ICBM's in silos, towards the strategic (and hence political) instability that would be the consequence were Soviet leaders to believe that they could coerce the West, either through intimidation in peace-time and crisis-time, or through actual military action on a limited scale in time of war.

The Report stresses again and again the need for the United States to "be able to put at risk those types of Soviet targets -- including hardened ones such as military command bunkers and facilities, missile silos and other storage, and the rest -- which the

Soviet leaders have given every indication by their actions they value most, and which constitutes their tools of control and power."[9] The Report goes on to argue that "[a] one-sided strategic condition in which the Soviet Union could effectively destroy the whole range of strategic targets in the United States, but we could not effectively destroy a similar range of targets in the Soviet Union, would be extremely unstable over the long run."[10] Hence, for the quality of deterrence mandated by the overseas security commitments of the United States, vis-à-vis a Soviet state not known to be overly squeamish over prospective loss of life among its civilian population, the distinctive attributes of an ICBM force are essential. Denying any crumb of comfort to those still attracted by counter-city, assured destruction reasoning, the Report concludes its discussion of deterrence with the flat assertion that:

> ...the deterrent effect of our strategic forces is not something separate and apart from the ability of those forces to be used against the tools by which the Soviet leaders maintain their power. Deterrence, on the contrary, requires military effectiveness.[11]

Given what the Commission has to say later on about the uncertainties of communications with submarines and about the accuracy of submarine-launched ballistic missiles, it must follow that the United States has no responsible policy choice other than to modernize its ICBM force.

Peacekeeper and Its Basing

The Scowcroft Commission Report was compelled to tread a very narrow line between endorsing the traditional verities of crisis stability, which have -- as their centerpiece -- the axiom that forces should not be so deployed as to invite preemptive attack, and stressing the deterrent, and even stability, merits of forces which are not independently survivable. In essence, the independent survivability of ICBMs is desirable, but not truly essential (so long, that is, as the Soviet Union lacks an SLBM force sufficiently accurate as to pose a prompt threat simultaneously to missile silos and to bomber-bases, and so long as there is no serious doubt concerning the invulnerability of submarines). The President's Commission argued that "...whereas it is highly desirable that a component of the strategic forces be survivable when it is viewed separately, it makes a major contribution to deterrence even if its survivability depends in substantial measure on the existence of one of the other components of the force."[12]

Independent survivability traditionally has been an attribute of the ICBM force, and it has been the endeavor to continue this feature in the context of Peacekeeper deployment that has so convulsed defense politics since the late 1970s. In the Commission's judgment the silo-housing of 100 Peacekeeper missiles will be a "good enough" solution to the problem of providing the prompt hard-target counterforce capability that the United States needs for deterrence or, possibly, for escalation control.

On balance, and it is on balance, the Commission is surely correct. With reference to day-in, day-out diplomacy and to most phases of crises, silo-housing for the Peacekeeper will be quite good enough. A Peacekeeper-armed United States will be a United States that has effected an appropriate competitive response to the counterforce challenge posed by Soviet fourth-generation ICBMs. A United States so armed will pose a credible prompt threat to many, though far from all, of the most valuable assets of the Soviet state. This will be a more dangerous United States, in Soviet eyes, which will be healthy for deterrence.

Politically, as Commission members have stressed, it would be very damaging were the United States to "fail the course" on Peacekeeper deployment. Notwithstanding the excellent military and arms control rationales for a modest scale of Peacekeeper deployment, by far the strongest argument lies in the realm of perceptions by the Soviets and our allies. Four presidents have endorsed the Peacekeeper missile as essential to American security. Moreover, the deterrence arguments for the Peacekeeper specifically are so strong given the well-advertised direction in American strategic policy (PD59 of July 1980, for a leading example), that a failure to field this system unambiguously would be, and would be seen abroad to be, a failure of the American political system to do that which it acknowledged to be strategically necessary.

Strong though the case is for Peacekeeper basing in silos, it must not be forgotten that it is a politically coerced choice. There are good reasons why it is desirable, if not absolutely essential, for the ICBM force to be independently survivable. It cannot be denied that there is a tension between the argument of the Commission on the one hand that Peacekeeper deployment is needed urgently because of the value for deterrence stability of the potency of the threat that it will pose, while on the other hand silo-basing will be "good enough" for reason of the survivability that inheres in participation in the triad as a whole. The stronger the Administration makes its argument for the

deterrence value of the threat posed by Peacekeeper,
the stronger it has to acknowledge the Soviet incentive
to neutralize that threat. Notwithstanding the large
and different threats posed by the bomber and SSBN
forces it is, in principle, undesirable that prospectively (for the late 1980s and 1990s) the most militarily effective American strategic weapon should not
be able to ride out an attack.

The Commission has not sought to deny the validity
of this concern, though it has -- rightly -- stressed
the extreme nature of the scenario wherein it may be
relevant. Theoretically vulnerable basing in silos
could prove to be fatal for crisis stability only in a
situation where the Soviet Union was willing to launch
a massive attack against the ICBM fields, thereby
giving tactical warning for U.S. manned bombers and
cruise missile carriers to take off, and assuming that
a U.S. President (or a successor NCA) would not, or
would not be able to, launch ICBMs on warning or under
attack. Should the Soviet Union attack U.S. bomber
bases first, the Commission Report notes that launch of
the Peacekeeper-Minuteman force would be a case of
"launch after attack -- launch after massive nuclear
detonations had already occurred on U.S. soil."[13]

Overall, one must grant that the case against
silo-basing the Peacekeeper rests upon a long-odds
scenario -- a context wherein a Soviet leadership might
choose to escalate to central war, out of a theater war
because: (1) of the quality of threat it confronted in
the Peacekeeper force; (2) in theory the Peacekeeper
force could be neutralized; and (3) they would reason
that the United States would have to launch the Peacekeeper very promptly for fear of losing it.

Although long-odds scenarios do turn up, the odds
in this case can be lengthened still further in the
U.S. favor. Above all else perhaps, the United States
can increase the technical plausibility of launch
under, and after, attack for the silo-housed Peacekeeper force. It may be very important that Soviet
leaders not believe that their SLBMs could pin down the
Peacekeeper, or disrupt its command and control,
pending arrival of the hard-target killing ICBM salvos.
In addition, it is far from obvious that silo-housing
of the Peacekeeper is incompatible with independent
survivability. The technical promise of silo-
superhardening almost certainly is far greater than the
Commission Report suggests. Silos 25 times more resistant than at present may be possible by the late
1980s.

It is probably fair to say that there really is no
controversy over the silo-housing of the Peacekeeper.
Everybody would prefer an independently survivable
basing mode, but those preferences are so diverse that

they have not met, and now almost certainly cannot meet, to construct a politically viable consensus. In short, silo-housing is nobody's first choice, but it is the only basing story that is politically workable and it will enable the United States to field the missile that is needed, and that is available, to enhance the stability of deterrence through the remainder of this century.

The Small ICBM

By way of demonstrating its endorsement of the goal of survivability -- in part to compensate for the silo-housing proposal for the Peacekeeper -- the Commission recommended a development program for a small, single-warhead ICBM. Commission members have had to be careful lest they paint so attractive a picture of the merits of the small missile, that it be used as an excuse for the near-term abortion of the Peacekeeper. Such a small missile would have many attractive features in comparison with the Peacekeeper. Its small size and weight would facilitate agile deployment, and its single warhead would render the missile a relatively low value target. According to classic criteria of stability, the proliferation of small ICBMs should be an important gain.

However, attractive though the idea of the small missile is, the Commission is understandably nervous lest the idea be asked to bear more traffic than will be possible. The Commission Report is very careful not to suggest that Peacekeeper is an interim solution -- though housing in non-survivable Minuteman silos may be only an interim measure -- pending availability of the small missile. On the contrary, the Commission said that it "would not insist on seeking a single solution to all the problems -- near-term and long-term -- with which the ICBM force must cope."[14] Plainly, the Report looks to an ICBM mix of Peacekeeper and the small missile, and perhaps some residual Minuteman missiles as a hybrid solution to the ICBM survivability and target coverage/deterrence problems.

The Commission, and the Administration, is right to insist upon the Peacekeeper, not only because of the powerful political and perceptual reasons cited earlier, but also because the U.S. ICBM inventory should contain a force of missiles with a throwweight sufficiently large as to permit high-yield warheads, provision of penetration aids against the far-from-trivial possibility of Soviet deployment of much more extensive active missile defense, and the delivery of space assets into orbit.

Without denying the attractiveness of the small ICBM, there is a host of reasons why its prospective development should not be permitted to have an adverse impact upon Peacekeeper deployment. First, the small missile, at present, is a "vu-graph system." Opinions differ as to the rapidity with which the missile can be developed. While there would seem to be no high technical risks associated with the idea, the fact remains that a strategically significant number of small ICBMs could not be deployed before the period of 1993-95 (assuming an initial operational capability in 1992).

Second, although newspaper readers may be excused the belief that the Commission Representative, Albert Gore, Jr., and Henry Kissinger have just discovered the idea of the small ICBM, the fact is that this idea has been studied, off and on, for nearly twenty years. Although it is correct to assert, in principle, that small size and low weight facilitates agility in deployment, questions remain. Just how should a smaller ICBM be deployed? What would it cost? The kind of truck-borne mobility to which a small missile would lend itself in technical terms, plainly is a political impossibility because of public interface. Paradoxically, given the advantages over the Peacekeeper of ease of handling, it is at least arguable that the basing problems of a small missile could prove to be no more tractable than has been the case with the larger missile. There is no cheap and simple way to deploy and operate a force of many hundreds of small ICBMs. Suggestions to the effect that the small missile might be deployed in very heavily armored vehicles for mobility on military reservations are simply suggestions until the feasibility of the hard mobile launcher is proven.

One need not be a prophet of gloom to foresee yet another major debate over ICBM basing. A mobile small missile may be vulnerable to barrage attack and its communication links and its accuracy could both be affected adversely by movement. Casual talk of "mixed basing" for the small missile, far from solving the difficulties, rather may serve to compound them. The point here is not to say that survivable basing modes cannot be found for the small missile, but it is to say that on the basis of the Peacekeeper experience it is difficult to be optimistic. Any critic of the Peacekeeper who claims to have found in the small missile the solution to crisis stability problems, should be compelled to think through the prospective basing story for his preferred system.

Third, while there are no inexpensive ways in which the ICBM inventory can be modernized, it is very likely indeed that a small missile that was deployed in a highly survivable manner would be an extraordinarily expensive way to provide ready ICBM warheads. Housing only one warhead on a missile means the purchase of one launcher per warhead, one guidance system per warhead,

one armored truck per warhead...and so on. It is too
early to offer dollar estimates, but it seems safe to
say that a small missile worth purchasing on strategic
grounds -- that is to say, one that was truly mobile,
or one that was very heavily defended by concrete and
steel -- almost inevitably will be a small missile that
will be in deep trouble on budgetary grounds.

Fourth, while the small ICBM could offer an arms
control opportunity for both sides to de-MIRV their
forces eventually, this likely will not happen. It is
probable that the Soviet Union will decline any invitation to effect a START-licensed restructuring of its
ICBM force. The Soviet Union is not known to share
American fears of crisis instability; almost certainly
judges its very substantial force of heavily-MIRVed
SS-18s and SS-19s as a very efficient means of packag-
ing an opening counterforce punch; and already has a
potential mobile, small single-warhead ICBM, in the
form of the SS-20 IRBM (if two warheads are offloaded)
and is testing another ICBM, the SS-X-25, that appears
to be optimized for mobile basing. Moreover, the
verification problems associated with a small, genuinely mobile ICBM may not lend themselves to easy solution
in the predictable absence of "on-site" counting procedures.

Notwithstanding the negative points registered
above, it is necessary to agree with the Commission in
noting the synergism between the Peacekeeper and the
small single warhead ICBM. The threat to many Soviet
ICBMs in silos that will be posed by the Peacekeeper
may, indeed logically should, encourage Soviet interest
in deploying a more survivable and less threatening
ICBM force of their own. It is arguable, at least,
that unless the Soviet Union can be persuaded by the
Peacekeeper and (later) by the Trident D-5 threats to
draw down the quantity of its ICBM payload, it is
virtually impossible for the United States to deploy a
genuinely survivable small, single-warhead ICBM at
bearable cost.

Conclusions: Panaceas and Sound Programs

It is evident that the elements for a national
consensus on strategic force modernization and strategic arms control have been assembled. It is no less
evident that, as the Commission maintained very strongly, no single element in the package can stand on its
own. Deployment of the Peacekeeper ICBM must be the
first priority in the ICBM and arms control story,
because it is the basis for everything else. In the
absence of a visibly healthy Peacekeeper deployment
program, it is inconceivable that the Soviet Union will
be interested in drawing down the payload of their ICBM

force -- at least during the 1980s. Also, a U.S. ICBM force comprising, eventually, several hundred small, single-warhead missiles and perhaps a very modest-size force of very old Minuteman IIIs would lack the throw-weight needed for payload flexibility in the face of what certainly would continue to be very stressful threats. Furthermore, as of 1984, one cannot be certain that a small ICBM will prove to be politically feasible, once the full scope of the costs and deployment difficulties involved come to be appreciated widely and in detail. It follows that the United States prudently cannot forego Peacekeeper deployment, for in that unhappy event there may prove to be no modern ICBMs of any kind in the arsenal in the 1990s.

The small ICBM is desirable as a valuable element in a timely shift towards a more stable strategic force posture. Properly, that is to say survivably, deployed, it should lend useful assistance for deterrence to a Peacekeeper force housed in silos. The overall value of a counter-Peacekeeper strike would be reduced to an important degree were Soviet targeteers to confront, simultaneously, the difficulty of finding, tracking and striking a truly mobile force of small, but hard-target capable, single-warhead ICBMs.

NOTES

1. In discussions about the warning time available after detection of a Soviet SLBM launch against U.S. bombers and ICBMs, it has become fashionable to describe the "worst case": an SLBM launched on a "depressed" trajectory from a submarine about two to three hundred miles off the U.S. coast. Two things must be said about this "worst case": 1) in spite of its apparent attractiveness, the United States has not observed a Soviet SLBM flight test with a "depressed" trajectory (in fact, during the SALT II negotiations both U.S. and Soviet technical representatives found it so difficult to define a depressed trajectory for verification purposes that the subject was eventually dropped from the discussions); and (2) The majority of the Soviet SSBN force, in apparent deference to U.S. and allies ASW capabilities, appears to prefer the safety of patrol areas on the immediate periphery of the Soviet Union. Only

a few Soviet submarines regularly patrol in forward deployment areas.

2. During periods of normal day-to-day alert, over half of the U.S. SSBN force is at sea: either steaming to and from its patrol areas or in its SIOP-assigned patrol areas, available for SIOP missions. Those submarines that are steaming conceivably could be turned around after provisioning or reassignment.

3. For example, the United States is now replacing the solid rocket motors in its Minuteman II ICBMs. Also, there have been reports in the past that there were problems with certain elements of the Polaris SLBM reentry systems.

4. There is now some discussion about whether or not the "Stealth" bomber, with its somewhat limited payload capability and its $34 billion price tag, would be more cost effective than an additional 100 B-1C aircraft, designed with some "Stealth" features.

5. SLBMs and bombers each comprise about 20% of the day-to-day alert weapons in the U.S. strategic arsenal. Under generated conditions, each triad element would contribute about 33% of the alert weapons.

6. Report of the President's Commission on Strategic Forces, (Washington, D.C.: Department of Defense, April 6, 1983), p. 9.

7. Ibid., p. 25.

8. Ibid., p. 12.

9. Ibid., p. 6.

10. Ibid., p. 6.

11. Ibid., p. 11.

12. Ibid., p. 8.

13. Ibid., p. 13.

14. Ibid., p. 14.

3
Technology Impacts on ICBM Modernization: Hard Mobile Launchers and Deep Basing

Blair Stewart

Since the early days of the MX program, new technological developments have had an important impact on the course of the U.S. ICBM modernization program. Indeed, the controversy over ballistic missile accuracy and the resulting implications for the survivability of fixed ICBM silos has been the dominant theme in the decade of debate over the nature and character of new U.S. ICBM developments.

Clearly, the question of technological feasibility still remains a major issue in the U.S. search for a survivable basing mode for all or a part of its ICBM force. It is therefore appropriate to examine the most promising technological developments for enhancing ICBM survivability and their potential impacts on the strategic balance, arms control, and the U.S. ICBM modernization program in general.

Likely Technological Developments

At present, there are several developments which appear to offer varying degrees of promise to improve the pre-launch survivability of the U.S. ICBM force: (1) a hard mobile launcher for the new small missile; (2) deep underground basing; and (3) superhardened silos for either the Peacekeeper or small missile or both. The first two will be examined in this chapter; superhardening will be examined in a later chapter.

Hardened Mobile Launchers

As recommended by the Scowcroft Commission, the Air Force is now developing a small ICBM of about 30,000 pounds gross weight, primarily intended for some form of mobile basing.[1] One of the possible mobile basing modes under consideration is a hardened mobile launcher, designed to withstand nuclear blast over-pressures on the order of 25-50 pounds per square inch (psi). Survivability would be achieved by forcing

Soviet targeteers, who cannot know the precise location of the launcher at the time of either SLBM or ICBM launch, to barrage attack a large area where the launchers may be located.

Theoretically, a hardened launcher would reduce the lethal radius or "kill area" of a nuclear weapon. An attacker would thus be forced to expend a larger number of weapons against hardened launchers in a given area to insure high kill probabilities. The principle advantage of a hardened mobile system is that it may allow a force of small missiles to be based and operated on existing military reservations, thereby reducing the problems associated with public interface -- an issue which has proved particularly troublesome to the MX program.

The major issue facing a hardened mobile launcher may be rather basic: the feasibility of such a launcher has never been demonstrated, although studies indicate that it may be possible. Key questions surrounding the feasibility of a hardened mobile launcher include: Can such a vehicle be constructed so that it can remain stable during a nuclear blast wave? Can the required hardness levels be achieved without securing the vehicle to the ground or some structure with restraining devices? Can the vehicle operate off-road to increase its available operating area, and can it operate in all types of weather? Can blast seals be designed and built which will not allow overpressure to leak under the vehicle, thereby causing it to overturn? What are the chemical, biological, and radiation protection requirements for the launch crew? Can the vehicle's electronic equipment be protected against radiation and electromagnetic pulse effects resulting from a nearby (one to two miles) nuclear detonation (the so-called source region EMP problem)? Can the vehicle be secured from terrorists and enemy agent threats, and is there an acceptable scheme for nuclear surety?

Deep Underground Basing

Some form of deep underground basing also may prove feasible for enhancing ICBM survivability. Key technology questions for deep basing include: Can a deeply buried complex large enough to house the required number of ICBMs be constructed? Can tunnel boring and mining technology be developed which would reduce the costs of construction to acceptable levels? Can reliable, secure, and survivable command, control, and communications be provided between the facility and the National Command Authority and/or higher headquarters? What are the requirements for crew support and living quarters? Can chemical and biological protection be provided for personnel? Can reliable and rapid excava-

tion be achieved in fractured rock? Can remote control
of excavators and launchers be accomplished? Can some
method of quick egress be provided for at least a
portion of the ICBMs to insure a rapid response capability against time urgent targets?

How these technological questions are resolved
will have important impacts on the direction of U.S.
ICBM modernization efforts. While it is important to
delineate these technology issues, it is not the
purpose of this chapter to evaluate prospects for
deriving acceptable answers and solutions for all of
them. Rather, this analysis looks at how mobile hard
launchers and deep underground basing might affect the
U.S. choices of future ICBM basing modes. This
requires a review of the following aspects of U.S.
ICBM basing alternatives: political acceptability,
potential environmental impacts, relative costs, operational implications, arms control implications,
possible Soviet reactions and strategic considerations.

Hard Mobile Launchers

Political Issues

Given the current euphoria in Congress over small
missiles and the perceived wonders this missile can
work for strategic stability (and barring any major
technical "showstoppers"), a force of small missiles on
hardened mobile launchers operating on existing military reservations would appear to be the most politically acceptable ICBM modernization option currently
available to U.S. decisionmakers. But the previously
mentioned technical issues surrounding this system may
have significant impacts on its political acceptability.

First, the tradeoff between vehicle hardness and
the amount of land area required to insure survivability may have a critical political impact. At first
glance, the U.S. military would appear to have more
than enough land on which it conducts its normal operations and training to allow for the CONUS basing of a
force of small, mobile ICBMs (for example, the six
largest DoD reservations in the continental U.S.
encompass a total area of over 18,000 square miles).
But this apparent abundance of operating space dwindles
to some 12,000 square miles when this land is matched
against the technical criteria required for acceptable
mobile operations.

For example, there must be sufficient land with
proper grades (less than about a 10 percent slope) to
allow the launcher vehicles to traverse the land without too much difficulty. Additionally, the existence

of improved and unimproved roads will affect the costs to deploy such a system and the effective land area in which the vehicles might be located.[2]

Finally, there is the question of how operation of a mobile ICBM force might impact current missions on these military reservations. The best example of this situation is the Nellis AFB gunnery and bombing range located in Nevada. This is a rather large land area (about 5,000 square miles) exclusively devoted to military use. But the missions it supports (including the adjacent Nevada Test Site for DOE weapons testing) simply cannot be performed effectively anywhere else in the United States.

Consequently, decision-makers will be faced with a difficult tradeoff. Indeed, since this system will require thousands of square miles of land area to achieve acceptable survivability, finding enough suitable land will not be an easy task.

Given these land usage constraints, the political acceptability of a small mobile missile in all probability will be closely related to the degree of public interface required for its operation. If sufficient hardness can be built into the vehicle to allow it to operate in a smaller land area, and if sufficient, technically suitable land can be found on existing military reservations, this should improve the political acceptability of a mobile ICBM. On the other hand, if these conditions cannot be met and the vehicles must operate on land outside of existing military reservations, this most assuredly will increase the concerns of potentially affected citizens and, hence, the anxieties of politicians who must answer to them.[3]

Additionally, one cannot overemphasize the potential political impact of how the American public will view the technical issues associated with hard mobile basing. It can be argued strongly that one of the nails in the coffin for the MX/MPS system was the public's perception that the concept was a "Rube Goldberg" product. Defense professionals may have understood how the MPS system would enhance ICBM survivability, but the man on the street was unconvinced (perhaps baffled) and therefore skeptical of the concept. Eventually, opponents of the MX successfully played on this condition to create a fatal level of doubt about, and opposition to, MPS.

Clearly, this situation continues today and it could come to plague the small ICBM. If the techniques required to achieve the necessary hardness levels for a mobile launcher become very sophisticated and very complicated, organized anti-nuclear modernization groups can again be expected to raise the spector of another "Rube Goldberg" basing concept to generate additional opposition to the small mobile ICBMs.

Finally, acquisition costs of major weapon systems are always a political issue. Whether or not costs become a significant factor impacting the small ICBM will depend on how well the Air Force can solve the major technical and basing issues within cost boundaries which are perceived by the political process to be "reasonable." For example, a mobile system which has very large manpower requirements and hence, very high operations and support costs, may experience considerable opposition during the annual Congressional budget process.[4]

Environmental Impacts

Assuming that a hard mobile launcher (HML) system is based in the Southwest U.S., environmentalists may oppose HML because: (1) a certain amount of permanently disturbed land will result from the construction and upgrading of several thousand miles of roads and the construction of facilities for operations and support activities and personnel housing; (2) there might be "unacceptable" socioeconomic impacts resulting from an influx of both several thousand construction personnel and a permanent operations and support party into rather sparsely populated areas; (3) there would be an increased demand on water resources; and (4) there could be negative biological and archaeological impacts in the proposed operating areas. These potential environmental impacts are not unlike those which surrounded MPS, although they would probably be on a much smaller scale and could not be challenged as easily in the courts -- particularly if the system is to be based on existing military reservations.

Operational Implications

A mobile ICBM which has hard target kill capability should prove quite troublesome to Soviet targeteers. A system of this type could severely stress Soviet nuclear battle management and strategic operational planning capabilities, and as a result, force the Soviets into some rather difficult and costly development efforts for effective counters. Possible Soviet countering options include improved satellite reconnaissance capabilities and increased intelligence agent activities to provide near real-time location data on the mobile launchers, and advanced "smart" reentry vehicles which could home in on the launchers after ascertaining a final target position fix while inbound to the target area.

From the U.S. standpoint, how much of a deterrent the Soviets perceive a mobile ICBM to be will depend on how effective the system can be made to appear. For example, if the system is advertised as a hardened mobile launcher, then its hardness or hardness characteristics must be clearly demonstrated in a manner such that the Soviets understand its survival capabilities and the magnitude of the effort required to neutralize the system.

Likewise, the U.S. must be able to solve the problems of accuracy associated with mobile systems and the difficult command, control, and communications problems attendant with a mobile ICBM in order to create a system which is truly perceived by the Soviet General Staff as both robust and potent.

Additional issues are those of warning dependence and endurance which will affect the operational effectiveness of a hardened mobile system. It is possible that the final system configuration will require some amount of advanced warning for dispersal of the mobile launchers in order to increase their pre-launch survivability. Although in the past the U.S. ICBM force has enjoyed warning independence, it is clear that the responsiveness and operational effectiveness of today's strategic offensive systems are forcing a higher degree of warning dependence on U.S. strategic forces. Consequently, warning dependence may already be a fact of life for U.S. ICBMs, and this in turn places renewed emphasis on the upgrading of U.S. warning systems such as the Tactical Warning/Attack Assessment system.[5] The point simply is that the warning dependence of a mobile ICBM system is not necessarily a "showstopper," as long as equal attention is paid to the necessary improvements in complementary U.S. warning systems.

Finally, mobile systems by their very nature are not known for their endurance in wartime. This condition could only be heightened in a nuclear war. Not to denigrate either the ingenuity of the U.S. technical community or the determination of SAC maintenance personnel, it will prove somewhat of a surprise if a mobile ICBM system can be conceived with a realistic concept for endurance beyond thirty to sixty days.

Possible Soviet Reactions

As a general rule, the initial Soviet response to a hard mobile system will be the same as it is for any other potentially threatening weapon system: efforts to prevent its construction or to delay it and reduce its operational capabilities such as their propaganda barrage and diplomacy directed against the current NATO theater nuclear modernization.

It appears relatively easy for the Soviet Union to influence the U.S. weapons acquisition process. Through statements, arms reduction proposals, and other means of playing the various U.S. and NATO special interests against one another the Soviets can influence U.S. weapons acquisition and deployment. At present it appears that Soviet attempts to influence the small

ICBM debate have become a secondary priority to their opposition to deployment of Pershing IIs and cruise missiles in Europe. But once the small ICBM program reaches a point of high public and political visibility, we should expect the Soviets to pick up their campaign to influence the debate.

Having waged such a campaign, the Soviets will then make their own acquisition decisions to develop the capability to counter a U.S. hard mobile system.

Deep Underground Basing

Political Issues

The potential political issues surrounding a deep basing system have relatively little to do with the actual technical merits or demerits of such a system; rather, the major issue appears to be whether or not the United States needs an enduring capability for the conduct of nuclear war.

The Reagan Administration is well aware of the proverbial hornet's nest stirred up by the news media when it became public knowledge that U.S. defense planning was reviewing the dynamics of a nuclear war and the possibility that such a war might be prolonged rather than abruptly halted after an initial, presumably massive, exchange. Given that the nuclear freeze movement continues to find it profitable to advertise any nuclear war as "the end of the world," it will be difficult to gain popular support for the idea of a citadel of military weapons and personnel buried deep inside the earth to continue a nuclear conflict after a large portion of the population has been destroyed.

Articulation of the rationale for deep basing will trigger the raising of this emotional issue, and unless the public can be better educated on complex military matters and nuclear deterrence, the concept is unlikely to receive a fair trial.

Finally, it is difficult to imagine a deep basing system which would not entail rather substantial costs for development, production, and construction. Again, the potential political impacts of costs would depend on the degree to which the Air Force is able to solve the technical issues surrounding deep basing. In general, a deep basing system deploying 100 Peacekeeper missiles, or 500 to 1,000 small ICBMs, could have life cycle costs that are two to three times that of 100 superhard silos. The cost/benefit argument for such a system most likely would be improved if such a system also were intended to house a National Military Command Center (NMCC) and boosters for post-attack reconstitution of critical U.S. space assets.

Environmental Impacts
      An additional political liability of deep basing
may be its perceived environmental impacts. Most of
the twelve potential deep basing sites identified in
initial studies are located in sparsely populated, arid
regions of the United States. Depending on the size of
the missile force to be housed and whether or not the
facility would also serve as an NMCC, there could be as
many as fifteen to twenty thousand construction workers
required to build such a system, and about ten to
twelve thousand personnel required to operate and maintain it after construction.
      These numbers imply rather large socioeconomic
impacts on areas with minimal existing support capabilities, perhaps on the order of the impacts which
were projected for the MX/MPS system. Additionally,
construction will place high demands on water in the
proposed deployment areas with the attendant legal
battles which were looming on the horizon for MPS.
      Once a deep basing system had been constructed,
there would be relatively little public interface
associated with the missiles or their warheads. The
only time this would occur would be when the missiles
and warheads were moving to and from the deep-buried
launch facilities.

Operational Implications
      Operationally, a deep basing system would create
difficult targeting problems for the Soviet Union.
Soviet attack options would include large yield weapons
to dig out the system, direct attack of the multiple
egress portals (whose location would be very difficult
to conceal), earth penetrators, and delayed fuzing
weapons designed to detect missile egress activities.
      In all cases, the Soviets could not be certain
that they had destroyed the ICBMs housed in the deep
basing system. Consequently, the Soviet leadership
would face the potential prospect that at almost
anytime after the start of a nuclear war, the United
States would retain an effective capability to attack
and destroy high value targets in the Soviet
Union.[5] In this respect, a deep basing system might
offer the best alternative for an ICBM secure reserve
force.
      Additional benefits of a deep basing system
include its resilience to reactive Soviet threats and
its previously mentioned potential for use as an NMCC
and as a warehouse and launch facility for the
reconstitution of critical space assets.
      On the negative side, while a deep basing system
conceivably could be built with high survivability and
long endurance, it is not clear that this can be

achieved while preserving the traditional quick response attributes of present ICBMs. The egress process would involve time consuming tunneling and boring, and although there are concepts for inclusion of quick reaction portals in a deep basing system, these at present appear to be either very impractical or very expensive. One important question is what happens to a time-sensitive target during the interval (perhaps as much as several hours) in which a missile is digging its way out of the rock?

This consideration is less important if the ICBMs in a deep basing system are primarily designated as secure reserve force weapons and are not essential to any prompt U.S. response once a nuclear exchange begins. Nevertheless, it is an important consideration in deciding whether or not to deploy U.S. ICBMs in an underground base and when deciding what roles and missions to assign them.

The egress process would seem to present an additional operational problem, that of a high degree of remote control operation. While remote control operations certainly are an integral part of the current U.S. ICBM force, none involve the lengthy time periods for mechanical and electrical processes associated with missile egress, launcher stabilization on the surface, missile ejection, and first stage motor ignition. If some part of the tunnel boring machine, the missile, or its launch equipment were to fail during the digout process, it could be virtually impossible to repair.

Possible Soviet Reactions

Regardless of the operational responses adopted by the Soviet Union, their ability to destroy quickly a deep-based secure reserve force will remain uncertain. Because of this the Soviets are likely instead to explore ways to break the C^3 link to the National Command Authorities. The Soviets may seek to isolate the system, then methodically reduce it with large yield warheads, earth penetrators, and timed fuses.

Deep basing strikes at the heart of Soviet strategic doctrine. A secure reserve of land-based missiles would prevent the Soviets from achieving the decisive shift in the "correlation of forces" which is critical to their theory of victory. For this reason, the Soviets would certainly wage an aggressive political campaign to delay or prevent such a deployment. By combining dire warnings concerning the increased likelihood of war with questions about the verifiability of deep basing, the Soviets might expect to confuse and frighten public opinion into demanding accommodation.

Conclusions

As one might deduce from the preceding discussion, no single basing mode has the ability to satisfy all of the diverse and sometimes conflicting criteria that from time to time have been placed on U.S. ICBM modernization alternatives. It is clear the Scowcroft Commission recognized this dilemma and therefore recommended a "mixed bag" of modernization actions. Accordingly, U.S. decision-makers should keep this in mind as they deliberate on the various ICBM basing alternatives.

This may be particularly true for the small ICBM. On initial appraisal, a mobile ICBM would appear very attractive from both a deterrence standpoint and as means to reduce the superpowers' dependence on MIRVed ICBMs. But there is a significant possibility that in its attempt to remain "technically pure," the U.S. scientific and engineering community may impose extremely stringent if not impossible survivability constraints on a mobile ICBM (for example, a threat scenario for small missile system definition concept studies that forecasts a Soviet allocation of thousands of warheads against the system). This type of scenario may in turn dictate overpressure hardness levels for the mobile launcher which become impractical to achieve, or dictate that the vehicles operate in expanded deployment areas, thereby increasing the potential for public interface problems.

It is not argued here that this type of attack would be an implausible method for the Soviets to use in an attempt to neutralize a hard mobile system; rather, the point is that such a large expenditure of ballistic missile warheads combined with additional nuclear attacks designed to achieve other wartime objectives might have significant impacts on Soviet strategic offensive force execution -- and therefore should reduce the plausibility of this tactic. This parameter must be given the appropriate weight in the deliberations and technical tradeoffs for a hard mobile system.

At this same time, survivability constraints should not be allowed arbitrarily to rule out consideration of mobile concepts other than a hard mobile system based on existing Defense Department reservations. For example, a garrison road-mobile system consisting of a small missile and a relatively simple tractor trailer launcher which could be dispersed with warning would appear to have high deterrent value. The Soviets would have to realize that in time of crisis this system could be dispersed on the world's most extensive highway network.

Additionally, if such a system were light enough to be air transportable, it could present the Soviets with the added prospect that some portion of the U.S. ICBM force could be dispersed to other locations in the world, thereby further compounding Soviet targeting problems. Conceivably, these U.S. ICBMs could be relocated to any number of areas with runways large enough to accommodate U.S. C-141s or C-5s.

Of course, this notion ignores or overly simplifies the difficult political issue of negotiating and obtaining landing rights for aircraft loaded with nuclear-tipped missiles. But there would appear to be significant deterrent benefit from such a system even if its relocation capabilities were only demonstrated from time to time in training exercises. This could be done on military bases within the CONUS.

Finally, if the weight of a mobile launcher is not overly constrained by survivability requirements, and these launchers are based in small groups (perhaps five or six) at many existing military bases in the CONUS, they could be transported by air, when required, to a central support base for maintenance and repair.

The discussion above is not of a proposed alternative to the hard mobile launcher system now under study; rather, it is an attempt to illustrate again the idea that the small missile program may represent the last significant modernization feasible for the U.S. ICBM force. Consequently, all parties to the ongoing debate must recognize the potential implications of any unrealistic constraints -- whether they are motivated by arms control considerations or other factors -- which might be imposed in the early stages of this program's development.

Finally, two important considerations must remain in the forefront of the deliberations on U.S. ICBM modernization: capability and flexibility. Clearly, highly capable U.S. strategic forces are more likely to command the attention of the Soviet leadership should they contemplate a nuclear war. Accordingly, a modernized U.S. ICBM force should possess sufficient military capability to threaten the spectrum of assets most valued by the Soviet leadership.

Flexible ICBM systems are important because the U.S. is moving into an era in which there will be fewer operational "rounds" to achieve the same targeting goals. Growing weapon system costs, arms control, and an increased Soviet capability to neutralize U.S. strategic forces all imply that fewer delivery systems will be available in the future for the same missions should we fail to deter the Soviet Union. It is therefore essential to high quality deterrence that whatever ICBM arsenal it evolves, the United States has a high degree of operational and targeting flexibility built

into it -- and that plans, procedures, and capabilities
are constructed that would allow the United States to
maximize the leverage of its offensive forces in the
event of war.

Given the long history of the U.S. attempt to
modernize its ICBM force, it is doubtful that a single
survivable deployment mode can be derived for U.S.
ICBMs. There appear to be several technologies likely
to be available over the next decade which could
enhance the survivability of the U.S. ICBM force.
While U.S. technological prowess may be such that the
detailed technical issues surrounding each technology
alternative can be resolved, each of these alternatives
does have certain political, environmental, operational, and arms control implications. Accordingly,
the most likely solution may be the design of a
composite force which takes advantage of the positive
aspects of several alternative basing modes in order to
produce an effective ICBM contribution to the U.S.
deterrent posture. What must be recalled is that the
credibility of the Peacekeeper program was hurt by the
perception that a new "best" basing mode was attempted
every few months. Multiple basing options for the
small missile should be maintained for all the reasons
discussed above, and so that a sequence of "best" solutions is not presented to the public and Congress; anytime an individual basing option is established as the
basing system, it will be judged by standards of ade-
quacy that no individual basing mode is likely to
meet.

NOTES

1.   See Report of the President's Commission on Stra-
     tegic Forces, (Washington, D.C.: Department of
     Defense, April 1983).

2.   For example, if the vehicle cannot operate off-
     road (which is a distinct possibility), a mobile
     system then becomes a line target rather than an
     area target -- that is, the Soviets will know that
     at any given time the vehicle is located somewhere
     on a network of lines uniquely defined by the
     existing road network. Accordingly, the land area
     chosen for operation of such a system must have a
     large number of roads or be suitable for construc-
     tion of new roads in order to extract a high price
     from a Soviet ballistic missile attack and to pro-

vide the requisite survivability for the system.

3. One need only witness the public outcry over the MX/MPS system to understand that this problem is real and probably has not disappeared since the 1980-1982 MPS confrontations.

4. Some Air Force estimates run as high as 25-30,000 personnel required to operate and maintain a force of 500 small missiles in a mobile mode.

5. The Tactical Warning Attack Assessment system consists of: 1) early warning satellites for both ballistic missile attack assessment and nuclear detonation detection; 2) the Ballistic Missile Early Warning (BMEWS) radar system; 3) the Pave Paws radar for SLBM launch detection; 4) the PARCS radar at Grand Forks, N.D.; 5) the Cheyenne Mountain Command Center; and 6) various other radars and assessment systems.

4
U.S. National Security Policy Implications of Superhard ICBM Silos

Barry R. Schneider

Introduction

Breakthroughs in ICBM silo construction suggest that hardness levels over 25 times that of current U.S. silos may be achievable. If so, calculations about future U.S. ICBM survivability and vulnerability will have to be drastically revised. The famed ICBM "window of vulnerability" could be slammed shut by this revolution in passive defenses. The superhardening of U.S. ICBM silos could require important changes in:

(1) the mix of U.S. strategic offensive forces.

(2) U.S. plans for strategic defenses against ballistic missile attacks.

(3) U.S deterrence and targeting policy.

(4) U.S. arms control policy.

(5) Soviet offensive countermeasures.

(6) Soviet ICBM deployment choices.

In 1983 the U.S. Air Force and the Defense Nuclear Agency jointly sponsored a series of tests of scale-model structures that indicated that silos might be built that could successfully resist explosion over-pressures 25 times greater than those forces that could crush present ICBM silos.

The renewed interest in U.S. silo hardening was sparked in the early 1980s by U.S. intelligence uncertainties in estimates of Soviet silo hardness. Soviet silo designs differed somewhat from U.S. designs and tests were initiated to more precisely understand the blast resistance qualities of Soviet ICBM silos.

As DoD testimony before the Committees on the Armed Services of the House and Senate indicates, early efforts of the U.S. Silo Test Program "revealed that scale-model steel clad and steel reinforced concrete structures, with high steel content, were surprisingly resistant to airblast loading."[1]

A series of eight tests of superhard scale-model silo structures were conducted for Peacekeeper and small ICBMs in 1982 and 1983. In 1984 structural hardness testing must extend to larger size silos, longer duration airburst pulses, and must include ground motion from a simulated ground burst attack.[2]

Before U.S. authorities can predict with high confidence the survival probabilities of U.S. ICBMs in new superhardened silos they need to test more fully and to understand:

(1) cratering phenomena,

(2) the effectiveness of new shock isolation systems,

(3) the hardness of missile launcher support equipment, and

(4) the problems and possibilities of missile egress from silos buried under tons of debris.

Finally, a validation test of all integrated missile and launcher equipment must be done. The Air Force and Defense Nuclear Agency integrated hardness technology program will address these technical challenges in FY 1985, FY 1986, and FY 1987.

Superhardening and the Structure of U.S. Offensive Forces

U.S. ICBMs in superhard silos may have a very high probability of survival against a Soviet missile attack. If so, this fact would have important ramifications for the American offensive force composition and for the role U.S. ICBMs are asked to play in the Single Integrated Operating Plan.

With superhardening there would be less pressure to launch U.S. silo-based ICBMs on warning or while under attack. U.S. National Command Authorities might no longer face such a severe "use or lose" dilemma in crises if Minuteman, Peacekeeper or other future ICBMs are housed in silos many times harder than at present.

The central role of the U.S. ICBM has been questioned recently as its survivability was diminished by the Soviet deployment of SS-18s and SS-19s. U.S. ICBMs traditionally have been considered the most important component of the strategic triad of ICBM, SLBM, and strategic bombers for a number of reasons:

(1) Only the ICBM provides the United States the hard-target capability needed to hold Soviet leaders, their forces, and their hardened command bunkers at prompt risk.

(2) The alert rate, availability, and reliability of ICBMs is greater -- force wide -- than the air-breathing or submarine components of the U.S. strategic triad.

(3) The ICBMs, once constructed and deployed, are the cheapest component of the triad to operate and maintain.

(4) There is no Soviet defense now or in the immediate future capable of stopping the ICBM, once launched.

(5) Command, control, and communications linkages are more reliable and promptly responsive between the NCA and ICBMs than between U.S. leaders and other elements of the triad.

(6) ICBMs are rapidly retargetable and are those forces best designed to execute promptly and precisely a limited nuclear option selected by the U.S. President in a conflict.

(7) The presence of ICBMs forces the Soviet war-planner to allocate a large share of his weapons against them, and prevents him from maximizing his efforts vis-à-vis U.S. fleet ballistic missile submarines and SAC bombers. Furthermore, the ICBM component of U.S. forces causes acute attack coordination and timing problems for Soviet warplanning.

(8) The Soviet leadership especially respects the power of the ICBM; therefore U.S. ICBMs are especially effective deterrents to Soviet adventurism.

A major upgrade in U.S. ICBM survivability against a Soviet first strike -- as represented by the development of superhard silos -- could preserve the central role to be played by intercontinental ballistic missile forces into the 21st century when active strategic defenses may begin to catch up with strategic offenses.

The ICBM and the triad concept are definitely here to stay even if superhardening of silos is not possible, but the greater stand-alone survivability imparted to ICBMs by superhardening strengthens the argument for their continued deployment.

Superhardening Implications for U.S. Deterrence Policy

U.S. superhard ICBM silos can act in the future as shields which protect U.S. ICBMs from a Soviet first strike attack. As such, they could preserve the U.S. retaliatory capability against Soviet leadership bunkers, C^3I nodes, and strategic forces. This is especially important, for it is believed that U.S. capability against these target sets maximizes the U.S. ability to deter Soviet adventurism and minimizes Soviet willingness to initiate conflict.

The superhardening revolution cuts both ways. Superhard U.S. silos and structures can reduce Soviet incentives to strike at U.S. ICBMs, U.S. leaders, and at critical U.S. C^3I assets. On the other hand, superhard Soviet silos, bunkers, and structures can remove from prompt risk the target set the United States most needs to threaten in order to minimize the risk of war with the Soviet Union during intense international crises.

Soviet leaders climb to the top of their political system by pleasing those above them in the hierarchy. Conceivably, Politburo members might be as deterred by the prospect of a threat to Soviet cities and the Soviet population as by a more direct military threat to themselves and their instruments of power. However, Soviet rule is not marked by an evident concern for the well-being of the many peoples of the U.S.S.R. Rather it is a regime run by the elite for the benefit of that elite first and foremost.

It is this fact that led U.S. officials to conclude that U.S. targeting and deterrence policy should emphasize retaliatory threats that maximize the danger to lives of the Soviet high military and political command, and to its chief instruments of power -- the U.S.S.R. military forces, Soviet war industries, CPSU headquarters, and the communications links between the Soviet leadership and their military forces.

From this understanding came Presidential Directive 59 and the "countervailing" theory of deterrence (and targeting) that was developed during the waning days of the Carter Administration. This targeting/deterrence philosophy reportedly has been adopted by the Reagan Administration. The countervailing strategy provides the fundamental rationale behind the acquisition of improved U.S. time-urgent hard-target capability as represented by:

-- Minuteman III upgrades.

-- The Peacekeeper ICBM program.

-- The Trident II SLBM program.

-- The Small ICBM program.

Superhardening could protect leaders, C^3I assets, and ICBMs on both sides from the best weapons available today. This could lead to an asymmetrical situation where the Soviets could perhaps threaten the highest values of the U.S. leadership (the existence of a large fraction of the American people and their major urban-industrial areas) and where the highest values of the Soviet leadership (their lives, forces, and control) were not put at the same level of risk. The net effect of a superhardened world (on both sides) may be a less stable international regime than presently exists.

It is possible that the United States has put itself in a box whereby it has declared a deterrence policy that cannot be implemented very well by current or projected U.S. strategic forces. The Soviet hardening program, modest as it is in comparison to the potential revealed in the present U.S. silo test program, may negate the capability of the United States Government to pose the deterrent threats required by the countervailing doctrine of deterrence.

Should the Soviets acquire the capability to build superhard structures this could remove entire categories of important Soviet targets from the reach of U.S. nuclear forces. The United States then might be forced to reemphasize the threat it poses to softer urban-industrial targets in the Soviet Union and to link deterrence of war to such threats. Alternatively, the U.S. might develop an even greater accuracy for future U.S. ICBMs to offset Soviet hardening programs.

The current Achilles heel of U.S. strategic retaliatory forces lies in the vulnerability to surprise attack of (1) U.S. forces, (2) U.S. National Command Authorities and (3) the command, control, communications and intelligence (C^3I) linkages between them and the forces.

Some believe that the only possible Soviet means to a meaningful "victory" in a nuclear conflict might come if the Soviets could paralyze the brain (the NCA) or the nervous system (the C^3I links) of the U.S. force.

Superhard structures might help preserve both U.S. leaders and critical C^{3}I nodes -- thus guaranteeing a coherent U.S. retaliatory response should the U.S. be attacked. To the extent superhardening provides NCA/C^{3}I protection and to the degree Soviet military planners appreciate this fact, the Soviet leadership should be deterred from nuclear risktaking.

Superhard structures clearly could improve the survivability of certain critical strategic assets during a nuclear conflict. The President of the United States, the Secretary of Defense, the JCS, the CINCs (SAC, PAC, LANT, EUR) might retire during crises or conflict into a deeply buried and superhardened structure and conceivably might survive all but a direct hit.

Space reconstitution assets (boosters, satellites) might be protected inside superhardened structures in order to preserve and restore U.S. strategic connectivity in the event of a nuclear war. The Emergency Rocket Communications System (ERCS) which consists of missiles on launchers housed in silos at Whiteman AFB would be more secure if placed in superhard structures. The challenge in these areas would be to design equipment that could still function after undergoing severe vibrations and dislocations.

Other very early candidates for superhardening would be the Launch Control Centers (LCCs) connected to U.S. ICBM silo-launchers. Superhardening the LCCs could dramatically improve the survivability of these critical links to U.S. ICBMs.

Overall, superhard structures might serve as almost invulnerable storage vaults for such critical strategic assets as:

-- U.S. strategic reserve forces (e.g., ICBMs, SLBMs, ALCMs, GLCMs, SLCMs, IRBMs, MRBMs, and other nuclear weapons systems).

-- National Command Authorities.

-- Leadership supplies, communications equipment, code making/breaking equipment, hotlines, etc.

-- Computers and vital data banks relating to finance, public records, government files, recovery and contingency plans for reconstituting the government, economy, and military forces.

-- U.S. Government stocks of currency, gold reserves, silver reserves, medical supplies, foodstuffs, and plans and documents needed to run a wartime regime.

The superhardening of structures could serve to protect leaders, wartime forces, and the equipment and resources needed to sustain some kind of nuclear war effort, days and weeks into the conflict. Superhardening would not help an appreciably greater number of Americans to survive a nuclear war but it could promise that any wartime enemy would continue to feel a coordinated nuclear retaliation by the United States. That enhanced ability to mete out punishment, in a short or protracted nuclear conflict should serve to deter better the event itself.

Superhardening Implications for the Strategic Defense Transition

In his March 23, 1984, "Star Wars" speech, President Ronald Reagan started the United States on the path toward improved strategic defenses. President Reagan asked "Would it not be better to save lives than to avenge them?" The President acknowledged that developing and maintaining viable strategic defenses would be a long, difficult effort.[3]

It is possible that President Reagan's strategic defense initiative may one day be sidetracked and blocked by technical obstacles, reasons of cost, or by future political opposition. For this reason it would be wise if the United States used a building-block approach to its defensive transition, deploying systems that have stand-alone value should the remainder of the BMD program be terminated or seriously delayed.

The United States has focused most of its past efforts on developing ground-based ballistic missile defenses that focused on intercepting enemy reentry vehicles during the last minute of flight as they reentered the earth's atmosphere. The U.S. could have a considerable terminal phase BMD capability in the near term. What some of the President's scientific and policy advisors have suggested is that such ground-based BMD be deployed first to protect U.S. retaliatory capabilities (ICBMs, LCCs) and the more important C^3I links between U.S. strategic forces and the National Command Authority.

Point defenses of U.S. ICBMs and LCCs could improve the survivability of U.S. missiles at the launch end, although Soviet BMD improvements may require modifications in the front end of U.S. ICBMs and SLBMs to ensure penetration to targets.

To minimize the threat to international stability, the United States must maintain the integrity of its punitive retaliatory threat while its multi-layered defensive shield is put into place. Until ballistic missile defenses can be made highly effective the security of the United States and its allies will remain guarded by its ICBM, SLBM, and bomber forces.

Superhard silos guarding U.S. Peacekeeper and Minuteman ICBMs can provide the earliest real defensive improvement to the missile force, even before the first crude active ballistic missile defenses are put into place. The combination of silo superhardening and early point defenses of ICBM and LCC silos could result in a very significant improvement in ICBM survivability against a Soviet first strike attack.

U.S. ICBM silos 25 times more blast resistant than current silos (reportedly 2000 psi)[4] should greatly reduce the single shot or double shot probability of kill (SSPK or DSPK) that a warhead from an SS-18 or SS-19 can inflict. (Note Figure 4.1.)

If a superhardening program were combined with a 50 percent effective BMD program, then the individual SS-18 attacks per ICBM silo would be as shown in Figure 4.2.

One conclusion that falls out of this analysis is that silo superhardening by itself may be able to do more to protect the silo-based ICBM force than could a BMD system capable of destroying half the enemy reentry vehicles. A second conclusion is that the combination of superhard silos (passive defenses) and of ballistic missile (active defenses) could substantially improve the survivability of U.S. ICBMs at the launch end.

The figures used above are calculations based on the assumption that what kills ICBM silos is blast overpressure measured in pounds per square inch. Superhard silos can by themselves substantially reduce such kills.

However, what these calculations ignore is cratering as a kill mechanism. An ICBM may "survive" in a superhard silo but if it is upside down in a crater, it might as well be counted as destroyed. Ballistic missile defenses would be a useful adjunct to superhard silos in protecting U.S. ICBMs because they would intercept Soviet reentry vehicles before they could kill U.S. ICBMs through blast overpressure or cratering effects.

It is unlikely, but conceivable, that the Soviet Union might cooperate in a transition toward strategic defenses, and, if so, this could reduce the offensive pressure versus U.S. active and passive defenses.

Despite their signing the 1972 ABM Treaty, the Soviets have never given up their goal of "reliable protection of the homeland." Evidence of this can be found in their ballistic missile defenses around Moscow, their heavy BMD research and development efforts, their heavy air defense deployments, their formidable (by Western standards) civil defense

Figure 4.1
SS-18 Kill Probabilities Versus U.S. ICBM Silos
(Present Silos vs. Superhard Silos)

	SSPK	DSPK
Present Silos	.6124	.8498
Superhard Silos	.1048	.1986

Figure 4.2
SS-18 Kill Probabilities Versus U.S. ICBMs
(Effects of Superhardening and 50% effective BMD)

	SSPK	DSPK
Present Silos - no BMD	.6124	.8498
Superhard Silos - no BMD	.1048	.1986
Present Silos with BMD	.3062	.6124
Superhard Silos with BMD	.0524	.1048

efforts, and by the strengthening of their strategic defense military organization.

On the other hand, Soviet leaders are likely to be hesitant to relax their efforts in building offensive forces which have generated such influence and respect for Soviet power. This reluctance is likely to be reinforced by their innate conservatism and by the entrenched bureaucratic interests now in support of the present offense dominant Soviet posture. For example, the Soviet Strategic Rocket Forces and their bureaucratic allies will almost inevitably resist such a loss of their share of the military budget. For these reasons, the Soviet leaders are most likely to prefer a balanced but cautious approach to the defensive transition rather than a radical shift toward strategic defenses. Indeed, they might well elect to retain their current strategic posture loaded in favor of the offensive. They could then seek to offset U.S. active and passive defenses mainly through offensive countermeasures.

Indeed, there is no present indication that the Soviet leadership is ready to make such a radical shift from offensive to defensive emphasis. President Reagan's ultimate aim in the March 23, 1983, "Star Wars" speech likely is at odds with what the Soviets will decide to do and probably will be achieved, if at all, despite their competitive efforts. Soviet cooperation in scaling down offenses while building up defenses appears to be unlikely. Unilateral U.S. steps such as ICBM silo superhardening and active point defenses of ICBM silos will likely have to be effective without any help from U.S.-Soviet offensive arms reduction agreements (e.g., START).

U.S. Superhardening and Soviet Offensive Responses

Were the United States to develop and substitute new ICBM silos many times more resilient than the present ones, the Soviet military would be presented with far tougher problems in targeting U.S. ICBMs. Their technological responses might be to:

(1) work on ICBM accuracy improvements such as MARV and terminal updates;

(2) increase warhead yield;

(3) develop earth penetrating and other reentry vehicles to maximize cratering effects.

Accuracy Versus Superhardening

Every time the Soviet offensive strategic missiles improve in accuracy even a small amount, it would take large increases in U.S. ICBM silo hardness just to stay even. For example, in order for a given silo or bunker to retain its present probability of survival, its ability to resist blast overpressure (measured in psi) must improve 700 percent for every 100 percent improvement in accuracy.* Until the new superhardening techniques were discovered it appeared that technological progress was continually on the side of the offense. Defenses appeared doomed by the quantum leaps achieved in explosive yield, speed and range of weapons delivery, and by remarkable improvements in accuracy. The combined effect has been to make virtually every fixed target on the earth's surface vulnerable to sudden and decisive destruction.

Superhardening may offset many of the recent gains made by offensive weapons against ordinary missile silo targets. Increasing the hardness of the target by 2500 percent may make ICBM silos immune to destruction by all but direct hits by thermonuclear warheads. Until future ICBMs are deployed with terminal guidance systems and maneuvering reentry vehicles capable of making the last minute corrections, cratering effects rather than blast overpressure may be the dominant kill mechanism used to disable the ICBMs protected by super-hard silos.

---

* A 100 percent improvement in accuracy is considered to be a halving of the inaccuracy (CEP) of a given missile system.

Greater Yields Versus Superhardening

The Soviets might also increase the yield of their warheads as another measure useful in increasing the probability of kill against superhard U.S. ICBM silos. If, for example, 100 to 200 of the 308 SS-18 ICBMs were equipped with single 20 megaton warheads in order to target 100 Peacekeeper ICBMs in superhard silos* the single shot and double shot probability of kill from blast overpressure would increase dramatically. (Note Figure 4.3.)

Figure 4.3
Probability of Kill Versus Peacekeeper Silos[5]

	1 Mt SS-18 RV		20 Mt SS-18 RV	
	SSPK	DSPK	SSPK	DSPK
2000 psi Silo	.6124	.8498	1.000	1.000
Superhard Silo	.1048	.1986	.7039	.9123

Cratering Versus Superhardening

Finally, the Soviets might design their reentry vehicles, yields of warheads and fuzes so as to maximize the size and depth of craters created as a result of the attack.

Ground burst weapons create far larger craters than airbursts. Earth penetrating reentry vehicles that explode seconds or minutes after surface impact would expel even greater quantities of earth and rock.

The dominant kill mechanism against superhard silos is more likely to be cratering and silo displacement rather than crushing overpressure. If a targeted ICBM silo falls within the lip of the crater formed by the attack, the ICBM within its encapsulated silo-launchers may escape physical destruction but nevertheless may be rendered useless if the silo was tipped upside down or on its side.

The size of a crater formed by a nuclear explosion depends on such variables as:

(1) the height of burst of the weapons.

(2) the yield of the explosion.

(3) the type of soil or rock in the target area.

---

* This "defractionization" of the Soviet ICBM force would require them to reduce ICBM warhead totals by 900 to 800 warheads, a plus for the United States.

Generally the closer to the surface, or the deeper under the surface the explosion takes place, the greater the crater left by a nuclear detonation. Also, all other things being equal, "the volume of the crater is roughly proportional to the yield of the explosion."[6]

Finally, the size and shape of the crater left is a function of the nature of the excavated medium. Craters of different dimensions are produced by weapons of the same yield and height of burst depending upon whether the target area has wet soil, wet soft rock, dry soil, dry soft rock, wet hard rock, or dry hard rock.

As a recent DoD report to Congress has stated:

> Critical to the assessments or survivability improvements is our understanding of crater-related phenomena, including crater size, ground motions, and stress during crater formation, and the extent and characteristics of the debris or crater ejecta deposited around the crater. Furthermore, these crater-related environments, particularly the direct-induced ground motions and stresses, must be simulated during the large-scale system validation testing.[7]

Most of the earlier U.S. cratering calculations and estimates, prior to 1983, were based on U.S. tests conducted in the Pacific test range on atolls of wet sand and soft wet rock. This led to crater size predictions far larger than would be valid for dry soil and dry soft rock, the kind of geology found at Warren AFB where Peacekeeper ICBMs will be deployed in silos.

As Edgar Ulsamer recently reported in _Air Force Magazine_,

> ...a growing body of evidence from subscale testing and theoretical modeling suggests that the craters and overburden caused by nuclear detonations in some geologies are much smaller than previously assumed.

The smaller the area of the crater, the less likelihood that a silo would be inside the crater, and silos even barely outside of the "lip" of the crater will remain operational. A silo that is within the crater of a detonating weapon will be tilted so that the ICBM inside can't be launched. New design techniques, however, make it probable that up to forty degrees of tilt will not necessarily preclude a successful launch.[8]

54

Recent tests done on dry soil and dry soft rock
also indicate that previous predictions on the depth of
debris to be deposited over silos were too pessimistic
by a factor of two. Egress through the ejecta material
from the crater appears more feasible than before.

In summary, considerable Soviet testing of
countermeasures will be needed to defeat superhard
silos through either blast overpressure or cratering.
New U.S. civil engineering breakthroughs and new better
understanding of cratering in dry soil and dry soft
rock both point toward improved survivability for U.S.
silo-based ICBMs.

Superhardening and Soviet ICBM Basing Choices

The recent U.S. interest in silo hardening was
sparked by U.S. intelligence interest in and controversy over the hardness of new Soviet ICBM silos.
Part of the purpose of the U.S. Silo Test Program was
to get a better understanding of the blast resistance
of the new Soviet silo designs. This program has
yielded two results to date.

First, U.S. tests indicated that Soviet silos were
far more resilient when subjected to blast over-
pressures than formerly believed; far more resistant to
overpressure than current U.S. ICBM silos.

Second, the U.S. tests experimented with a series
of new silo structural designs that show extraordinary
potential for increasing the blast resistance of U.S.
ICBM silos -- perhaps by a factor of 2500 percent over
present designs.

The tests heretofore have been of scale model
silos and it has yet to be demonstrated conclusively
that we can design and build shock-isolation systems,
support equipment, ICBMS, and full scale sized silos
that can perform adequately together at such high
levels of blast overpressure. Nevertheless, it appears
that the U.S. Silo Test Program could yield superhard
silos of a hardness that matches or exceeds even the
best current Soviet silos.

However, it is likely that the Soviets already
have or will construct their own "superhard" silos.
Indeed, the Soviet incentives to develop effective ICBM
silos exceeds those of the United States since nearly
70 percent of the Soviet strategic nuclear weapons
force loadings are currently on silo-based ICBMs
whereas only one quarter of U.S. force loadings are on
ICBMs.

Thus, the silo superhardening revolution will cut
both ways. It will help greatly in protecting U.S.
ICBMs from Soviet ICBM attacks, and vice versa.

The Soviets could also be expected to apply the
new superhardening techniques to its leadership and

C^3 shelters. This means that an important class of very high value Soviet targets might be removed from direct and prompt risk due to superhard shelters. This has important and disturbing implications for those who believe that deterrence of war is best achieved by maintaining a "countervailing" targeting capability against the very targets superhardening techniques could protect. Soviet superhardening could remove from risk the targets that the U.S. countervailing doctrine of deterrence says should be made vulnerable if Soviet adventurism in acute international crises is to be best deterred.

If the Soviets have achieved, or will achieve, silo superhardness of the same quality projected for future U.S. silos, this fact should reinforce their present commitment to maintaining great numbers of silo-based ICBMs.

Indeed, superhard silo-based ICBMs might be seen as a practical alternative to deploying new mobile ICBMs now under development such as the SS-X-24.

It is likely that the SS-18, SS-19, and SS-17 ICBMs and LCCs would all be housed in such superhard silos as well as new ICBMs (and LCCs) now under development such as the SS-X-24 and SS-X-25.

It is already known that the Soviet political and military leadership maintains a series of very hard command bunkers throughout the Soviet Union to house:

-- the Politburo (and within it, the Defense Council)

-- the Main Military Council

-- the Ministry of Defense high command

-- the General Staff

-- CINC Warsaw Pact

-- CINC Far East

-- CINC Strategic Rocket Forces

-- CINC Air Forces

-- CINC Air Defense Forces

-- CINC Naval Forces

-- CINC Ground Forces

-- Top SSR Leaders

-- Other Top CPSU Leaders

Superhardening techniques could be applied to leadership bunkers and key C^3I headquarters to better preserve the fabric of Soviet leadership and control of Soviet forces in nuclear wartime conditions.

In general, the introduction of superhard ICBM silos, LCCs and leadership bunkers should reinforce not only the protective walls of each but also the status quo. With superhard ICBM silos the Soviets might be protected against future U.S. Peacekeeper and Trident II missiles without paying the costs of shifting the structure of present Soviet strategic forces. Those with vested interests in present Soviet force distributions should welcome the superhardening revolution.

Superhardening of ICBM silos, LCCs, and leadership bunkers should somewhat dampen future Soviet enthusiasm for:

(1) mobile ICBMs and IRBMs

(2) transfer of strategic resources to bombers.

(3) transfer of strategic resources of SSBNs.

(4) ballistic missile defense of ICBMs.

Conceivably all of the above may occur in addition to Soviet adoption of silo superhardening techniques but superhard silos should provide strong arguments for defenders of a status quo which features silo-based ICBMs as its centerpiece.

Superhardening and U.S.-Soviet Arms Control

Superhard U.S. ICBM silos could significantly reduce the lethality of Soviet ICBMs. They might negate or significantly reduce the present hard target capability of Soviet SS-18s, SS-19s and of future SS-24s and SS-25s. Passive U.S. defense might well offset the accuracy gains of fourth and fifth generation Soviet ICBMs. Because Soviet ICBMs would appear less threatening to U.S. silo-based ICBMs, there might be less need to put forward U.S. arms control and arms reduction proposals that drastically cut Soviet ICBM strength. Steel and concrete might provide the same crisis stability result as cutting Soviet counter-silo strength at the negotiating table. In other words, defensive U.S. technology might solve the ICBM vulnerability problem, leaving arms control negotiations such as START free to concentrate on other goals -- such as devising terms that would help head off countermeasures to superhardening (MARV, terminal guidance, earth penetrators, etc.) and terms which would improve international stability (e.g., confidence building measures) without major structural changes in U.S. and Soviet forces.

In this manner, superhardening of silos might permit arms control agreements that can achieve real reductions augmented by a series of stability improvement measures without running the arms negotiation process into the excess internal resistance generated within each Government whenever major budget reallocations are required by arms control agreements.

Looking ahead, arms control negotiation efforts might be directed toward stopping countermeasures to superhard U.S. ICBM silos. Unfortunately agreements which limited ballistic missile terminal guidance, maneuvering reentry vehicles, and other counter-silo capabilities such as earth penetrating RVs also would undercut the kind of U.S. force needed to threaten the highest values of the Soviet leadership -- which might in turn be protected by superhard structures. Crisis stability (via superhardening) might be purchased at a cost to U.S. deterrence leverage.

Like MIRV, maneuvering reentry vehicles (MARVs) will help the U.S. ICBM and SLBM forces to better penetrate present and future Soviet defenses by making U.S. reentry vehicles more accurate and also more difficult to track and destroy as they maneuver toward their targets. Soviet offensive and defensive countermeasures could make it difficult to deploy adequate counterforce capability to implement the countervailing U.S. threat required by PD59. MARV technology such as the Mark 600 program, together with Peacekeeper ICBM deployments, and Trident II SLBM forces should help offset Soviet deployments of:

(1) superhard silos.

(2) superhard command bunkers.

(3) superhard LCCs.

(4) SAM upgrade programs.

(5) terminal ballistic missile point defenses.

Conversely, a Soviet MARV program will also improve Soviet capability to penetrate future U.S. BMD improvements and could improve the accuracy of Soviet reentry vehicles against hard and superhard U.S. ICBM silos, LCCs, and leadership shelters. A MARV ban might alleviate this Soviet threat but at a cost to the U.S.

in being able to use MARV to help it implement its own preferred deterrence strategy. A two-sided MARV world also might retard the transition to a defense dominant world.

Conclusions

Scientific and technological innovations since World War II have tended to reinforce the advantage of the offense over the defense. That may be changing today as superhard structures and ballistic missile defenses appear more feasible.

In 1945 the fission bombs "Little Boy" and "Fat Man" which destroyed Nagasaki and Hiroshima were a thousand times more powerful than conventional bombs of the same weight. The first fusion weapons invented shortly after World War II were a thousand times more powerful than fission weapons. In less than three years (1945-1947) the United States succeeded in increasing explosive yields a million times beyond what had been previously possible.

This quantum leap in destructive power was mated with another quantum leap in the ability to deliver the weapons over long distances. The first intercontinental bombers carrying the first thermonuclear bombs made every fixed point on the earth's surface potentially vulnerable to attack in a matter of a few hours.

A third technological innovation which increased the dominance of offensive forces over defensive forces was the perfection of ballistic missiles with an intercontinental range. A fourth breakthrough was the multiple independently-targetable reentry vehicle (MIRV) missile. ICBMs and SLBMs provided a quantum jump in the speed of attack. Missiles launched in Eurasia could detonate 6000 miles away in North America in 30 minutes. MIRVs could overwhelm and saturate the primitive ballistic missile defenses of the 1960s. The advantages of explosive capability, rapid time to target, and numerical superiority all favored the offensive over defensive forces.

Finally during the 1960s and 1970s improved guidance systems now made a quantum leap in missile accuracy possible. Even silos hardened to withstand several thousand pounds per square inch of blast overpressure were no longer safe from attack. The revolution in silo superhardening may be the first step toward a better balance between strategic offenses and strategic defenses.

A short time ago, U.S. defense analysts were driven to the conclusion that all fixed targets on the earth's surface appeared within the destructive reach of Soviet ICBMs. A dangerous new era for U.S. silo-based ICBMs was opened in the late 1970s and early 1980s by the deployment of the Soviet SS-18 and SS-19 missiles. A short time ago many strategists saw silos as an inadequate answer to guaranteeing U.S. ICBM retaliatory capabilities in future wartime attack conditions.

However, most U.S. defense thinkers were unprepared for the kind of defensive breakthrough that super-hardening promises. If U.S. ICBM silos can indeed be hardened 25 times their current levels (some say 50 times), then the ICBM vulnerability problem can be made to go away for some time. This would make point defenses possible and could slam shut the ICBM "window of vulnerability" that was accepted as a given just a year or two ago.

Superhardening represents a breakthrough in defensive prowess. If the U.S. Silo Test Program succeeds as planned, U.S. strategic policies will require a thorough rethinking to stay abreast of the new technological advance on this front.

NOTES

1. U.S. Department of Defense, "ICBM Modernization Program Annual Progress Report to the Committees on the Armed Services of the Senate and House of Representatives," January 13, 1984, p. 30.

2. Ibid., p. 30.

3. Ronald Reagan, speech to a nationwide radio-television audience, March 23, 1983. See "President's Speech on Military Spending and a New Defense," New York Times, March 24, 1983, p. 20.

4. FY 1983 Department of Defense Appropriations Hearings, HAC, Part 9, Page 744, testimony of Richard DeLauer, Under Secretary of Defense for Research and Engineering.

5. See Fred A. Payne, "The Strategic Nuclear Balance: A New Measure," Survival, Vol. 20 (May/June 1977), p. 109. This calculation assumes the SS-18 has a CEP of .14 nm a .55 mt yield, and a reliability of 100%. Kill probabilities are based on the formula:

60

$$SSPK = 1-.05^{CMP/(H/16)^{2/3}}$$ (Single Shot Probability of Kill)

$$DSPK = 1-(1-Pk\ HPT)$$ (Double Shot Probability of Kill)

HPT = Hard Point Target

$$CMP = \frac{y2/3}{CEP^2}$$

y = Yield

CEP = accuracy in nm

H = psi hardness

For the purposes of this calculation a nominal hardness for U.S. ICBM silos is assumed to be 2000 psi. See Edgar Ulsamer, "The Prospect of Superhard Silos," Air Force Magazine, January 1984, p. 74. This figure was officially confirmed by Richard DeLauer, Under Secretary of Defense for Research and Engineering in testimony before the House Appropriations Committee, FY 1983 DoD Budget Hearings, Part 9, Page 744.

6.   Samuel Glasstone and Philip J. Dolan, eds., The Effects of Nuclear Weapons, Third Edition, (Washington, D.C., DoD and DoE, Government Printing Office, 1977), p. 59.

7.   Op. Cit., DoD Report, p. 33.

8.   Edgar Ulsamer, "The Prospect For Superhard Silos" Air Force Magazine, January 1984, p. 75.

5
The Strategic Defense Initiative and ICBM Modernization

Keith B. Payne

On March 23, 1983 President Reagan set into motion a process that has the potential to alter radically the traditional emphasis upon offensive forces in the U.S. strategic force posture. The President's televised speech mandated an examination of the technology to counter the threat from ballistic missiles.[1] Secretary of Defense Weinberger clarified the President's intent: to examine the potential for achieving a comprehensive defensive capability -- a "thoroughly reliable and total" defense against all strategic weapons.[2] Following the President's speech, National Security Study Directive 6-83 mandated two studies, one an evaluation of the policy implications of comprehensive ballistic missile defenses and the second study, an examination of the technology required to eliminate the threat posed by nuclear ballistic missiles.[3]

Commenting upon the findings of the Technology Study the President's Science Advisor George Keyworth noted, "We can now project the technology -- even though it hasn't been demonstrated yet -- to develop a defense system that could drastically reduce the threat of attack by nuclear weapons, not only today, but those that could reasonably be expected to be developed to counter such a defense system."[4] The Future Security Strategy Study on policy issues reportedly concluded that defenses "can increase stability," and that the U.S. should emphasize near-term BMD technology that could defend selected strategic force and C^3 assets.[5] These conclusions reflect some apparent, but unnecessary, friction between a defensive emphasis which is oriented toward near-term technology and limited defensive coverage, and a more sophisticated "defense in depth" designed to provide comprehensive coverage for urban/industrial and military assets. The U.S. defensive goal will have important implications for U.S. ICBM modernization in the future.

This is an examination of the strategic defense initiative (SDI) and its implications for U.S. strategic policy. Of particular interest are the potential implications of the SDI for the role of strategic offensive forces, stability, and U.S. deterrence policy.

Why SDI Is Important

A fundamental reassessment of the role of strategic defense in U.S. strategic policy could lead to very significant changes in the future of the U.S. strategic force posture. For example, some advocates of strategic defense imply that defenses can be considered as an alternative to offensive force modernization. It takes little imagination to foresee the argument that if the U.S. and Soviet Union are on their way toward rendering ballistic missiles "obsolete," why should the U.S. continue to fund potentially "obsolete" programs? There is likely to be strong pressure from some quarters to redirect resources away from offensive force modernization and toward the SDI if strategic defense is viewed as a benign substitute for offensive forces.

Second, if the ABM Treaty is revised or terminated the Soviet Union almost certainly will begin deployment of BMD (probably its ABM-X-3 system) for limited coverage of selected high value military, political, and military-industrial assets. This could endanger U.S. deterrence policy given the emphasis in that policy for holding hostage Soviet military, political, and military-industrial assets. As the Scowcroft Report noted:

> ...we must be able to put at risk those types of Soviet targets -- including hardened ones such as military command bunkers and facilities, missile silos, nuclear weapons and other storage, and the rest -- which the Soviet leaders have given every indication by their actions they value most, and which constitute their tools of control and power. We cannot afford the delusion that Soviet leaders -- human though they are and cautious though we hope they will be -- are going to be deterred by exactly the same concerns that would dissuade us.[6]

A Soviet BMD capability, which significantly degrades the U.S. capability to hold these Soviet assets at risk, could degrade the U.S. ability to deter. Improved Soviet strategic defenses could require even greater U.S. ICBM capability to maintain current target coverage.

The type of additional requirements for the U.S. ICBM program that will result from a Soviet defense initiative is, in part, reflected in a recent reported letter from Senator Wallop (R.-WY) to Secretary of Defense Weinberger, indicating that he would not support strategic ballistic missile modernization programs unless they included a plan for hardening against the effects of high-energy lasers.[7] This favors deployment of Peacekeeper ICBMs since the stringent weight considerations of the small ICBM program could be made much more difficult by the additional weight penalty of such booster hardening.

A nearer-term Soviet BMD deployment program could involve the SA-X-12. Evidently the SA-X-12 air defense system has two different interceptor missiles, one for air defense and another built and tested for BMD.[8] The SA-X-12 air defense system could be deployed within the existing, extensive Soviet air defense framework and provide a base for a rapid Soviet BMD "breakout" through the stockpiling and rapid deployment of a BMD-capable SA-X-12 interceptor. Dr. Robert Cooper, Director of DARPA recently testified to Congress that because of this "breakout" potential the Soviet Union, "could suddenly spring forward full-blown with a significant capability."[9]

The potential rapid deployment of hundreds of BMD-capable interceptors would likely present an extremely complicating factor in U.S. ballistic missile targeting and force-sizing considerations (even if in the near-term the interceptor only had limited BMD capabilities -- possibly requiring the U.S. to deploy additional penetration aids and missiles in order to maintain U.S. targeting capabilities). During a defensive transition the U.S. would probably need to enhance its ICBM capabilities rapidly to compensate for an initial Soviet advantage in BMD deployment. Yet, during this same period there will likely be pressure to redirect funding away from offensive systems (which Soviet BMD will be said to have rendered "obsolete") toward strategic defense. There will also be pressures to negotiate arms control ceilings on and cuts in strategic offensive forces to facilitate the defensive transition, and place a ceiling on offensive/defensive arms competition.

A defense initiative will also have an impact on U.S. ICBM basing considerations. Hardened military assets such as ICBM silos and selected C^3I facilities are the most likely to receive initial coverage from a U.S. BMD system, because such hardened point targets need not be defended "perfectly" and would benefit the most from the type of coverage that will be feasible in the near-to-mid-term.[10] In particular, a basing mode that employs some combination of deception and mobility, or closely-spaced basing and superhardening would lend itself to active defense because of the limited BMD effectiveness necessary, the limited threat corridor that must be defended, and the tremendous defensive leverage that might be available through preferential defense. Preferential defense (by a very limited number of BMD interceptors) of selected missiles or areas in an ICBM field or dispersal zone could double or triple the necessary number of attacking Soviet warheads in the absence of active defense.

Yet, an ICBM basing mode dependent upon active defense for survivability could become vulnerable to the political problems confronting BMD (such as the ABM Treaty). Consequently, if the SDI were short-lived for political reasons, any ICBM program closely related to BMD would also become vulnerable.

The Defensive Transition:
Toward a Limited or a Comprehensive Defense?

The Strategic Defense Initiative, according to early pronouncements by the Reagan Administration, has as its objective a "total" defense against nuclear weapons for the U.S. (and its allies). Yet, at least one of the policy studies stemming from NSSD 6-83 (the Hoffman study) emphasized the importance of more limited defenses for the U.S. strategic retaliatory potential.[11] In a recent report to Congress Dr. Richard DeLauer, Under Secretary of Defense for Research and Engineering, observed that, "The principal purpose of strategic defense is to enhance the survivability, and hence the effectiveness of strategic deterrence -- the National Command System network, strategic retaliatory forces and our military force and base infrastructure."[12]

Whether the goal of the SDI is seen as a limited defensive capability, or a comprehensive defense of U.S. urban and industrial assets is extremely important in terms of its impact on ICBM basing considerations.

For example, if a limited defense against light, counterforce, accidental, and/or third party attacks is the basic objective, then the development of passive protection for U.S. ICBMs and C^3I assets could, to some extent, mitigate the rationale for an SDI. The potential survivability for land-based military assets that may be attainable through super-hardening, deep underground basing, deceptive basing and/or mobility could prove to be adequate without active defense.[13] If passive defenses can be viewed as adequate for limited purposes, then the licensing of Soviet BMD deployment by U.S. withdrawal from the ABM Treaty would be seen as an unnecessary cost for the enhancement of ICBM silo survivability. This is an important point because if the U.S. does pursue a

policy that pushes the Soviet Union toward BMD, it will
be accepting the targeting degradation that Soviet BMD
would impose as a price worth bearing for the purpose
of actively defending American retaliatory forces.
Yet, if U.S. forces can be protected passively then an
SDI could entail the "price" of Soviet BMD deployment
without a corresponding useful increase in U.S. ICBM
pre-launch survivability over that provided by passive
defense. The net effect of this would be a likely
decrease in U.S. strategic effectiveness because of the
reduced access to Soviet high-value targets imposed by
Soviet BMD. If so, then deployment of limited ballistic missile defenses could work to the net disadvantage of the U.S.

Of course, the Soviet Union may pursue BMD deployment regardless of U.S. actions. The Soviet Union may
be preparing to "breakout" from the ABM Treaty and
deploy BMD systems independent of whether or not the
U.S. deploys strategic defenses.[14] If so, then a
U.S. SDI would not likely be to the net disadvantage of
the U.S. because the Soviet Union would be proceeding
with its defense in any event. Indeed, in such a case,
the U.S. could be in a much more disadvantageous
position if it pursued only passive defenses and
penetration aids for its strategic forces.

If the goal of the SDI is and remains, as identified by Secretary of Defense Weinberger, a "total"
defense against nuclear weapons, then no amount of
passive protection for military and/or civilian assets
will substitute for BMD. The critical issue associated
with a comprehensive area ballistic missile defensive
capability concerns the role that ICBMs would play in a
heavily defended world, a role that would be sure to be
questioned.

The SDI And the Role of Offensive Forces

It is possible that even if defensive technology
proves feasible, the transition to strategic defense
could be unacceptably dangerous.[15] Stability during a defensive transition would need to be maintained
by preserving the U.S. strategic retaliatory capability
in the face of improving Soviet defenses.

During a defensive transition there are likely to
be at least three destabilizing factors. First, the
Soviet Union may consider initial U.S. deployment of
BMD systems or components in space to be so inimical to
Soviet security that they may decide to disable
them.[16] There are two reasons why the Soviet Union
might attempt to deny the U.S. such access to space.
Soviet leaders would have incentives to attack a U.S.
space-based BMD component if it believed deployment of
U.S. strategic defense threatened to give the United
States a strategic advantage which the Soviet Union

could not redress. Attacking an unmanned spacecraft could be viewed as less provocative than virtually any other type of armed conflict with the United States. Thus, the risks could appear to be relatively low, while the benefit of denying the United States space-based defenses could be viewed as extremely important.

Second, the SDI could endanger stability because it is likely to lead to an initial Soviet advantage in deployed strategic defenses. The Soviet Union empha- sizes strategic defense to a much greater degree than does the U.S., and already possesses a formidable air defense capability; and unlike the U.S. the Soviet Union has made a serious commitment to civil defense. The Soviet Union also has pursued a BMD program that now presents the possibility of a relatively rapid "breakout" of BMD forces. As the Defense Department's annual edition of Soviet Military Power observes, "The Soviets have developed a rapidly deployable ABM system for which sites could be built in months instead of years.... The Soviets seem to have placed themselves in a position to field relatively quickly a nationwide ABM system should they decide to do so."[17]

If existing Soviet counterforce offensive forces could degrade the U.S. retaliatory potential, the com- bination of Soviet active and passive defenses, includ- ing BMD, might be considered adequate to absorb U.S. retaliation with acceptable losses. Such a condition would endanger strategic stability.

Third, at a more advanced phase of a defensive transition both sides may have achieved a capability to defend effectively against light attacks, or a retalia- tory attack by forces that had been degraded via pre- emption. In such a condition both sides could perceive powerful incentives to strike first. To delay striking first during a crisis could grant the opponent the opportunity to preempt. Because both sides' defenses could cope with a retaliatory strike, the disincentives against preemption could be less persuasive than the incentives to preempt. This type of instability has been discussed extensively, and often is labeled "crisis instability."[18] Crisis instability is thought to apply when a significant advantage would accrue to the side striking first, and when the disincentives to striking first are reduced because retaliatory threats are not effective. In such a con- dition powerful incentives would exist to strike first during a crisis because each side would fear that its deterrent would be ineffective, and because striking first could mean the difference between victory and defeat.

The U.S. has long taken the prospect of crisis instability seriously. As a consequence it has been cautious concerning the deployment of counterforce

offensive forces that could appear to threaten the
Soviet deterrent. As then Secretary of Defense Harold
Brown noted: "In the interests of stability we avoid
the capability of eliminating the other side's deterrent, insofar as we might be able to do so. In short
we must be quite willing -- as we have been for some
time -- to accept the principle of mutual deterrence,
and design our defense posture in light of that
principle."[19] The deployment of strategic defenses
by both sides, particularly if complemented by large
numbers of effective "counterforce" offensive forces
could engender the "crisis instability" discussed in
the above quotation.

These three potentially destabilizing factors
stemming from a defensive transition could be minimized
by the proper modernization of U.S. offensive forces.
For example, if the U.S. maintains modernized strategic
offensive forces, it is very unlikely that the Soviet
Union would start a war of attrition in space by pre-
empting initial U.S. deployments. The Soviet perspective on this issue is clear. War is a function of
political objectives, not technological developments.[20] The Soviet Union would be unlikely to
start a war with the U.S., even in space, if it were
unprepared for terrestrial conflict. If the Soviet
Union were prepared for central war, then it certainly
would target U.S. space-based assets, but it is
unlikely to commit an act of war against the U.S. in
the absence of such preparedness. As John Erickson has
observed, "Soviet military opinion cautions against any
'adventurist strategy' which might preemptively
initiate a total struggle when the requisite 'correlation of forces' (sootnoshenie sil) cannot assure a
favorable outcome (even 'victory') in such a
struggle."[21] Consequently, if the U.S. modernizes
its offensive strategic forces it should retain the
capability to deter an "adventurist strategy" such as a
preemptive attack on U.S. spacecraft.

Rapid Soviet BMD deployment leading to a unilateral defensive capability could (if complemented by
Soviet planning for preemption) possibly persuade
Soviet leaders that "victory" had become possible.
Yet, again modernization of U.S. offensive forces to
address the possibility of Soviet BMD breakout should
help ensure the maintenance of stability. Such modernization could include a variety of methods for nullify-
ing initial Soviet BMD deployments (decoys, chaff,
MARVs, cruise missiles, etc.). According to General
Forrest McCartney, former Commander of the Ballistic
Missile Office, Air Force Systems Command, there is
little doubt that the U.S. could retain a robust capability to target the Soviet Union even in the context
of initial Soviet BMD deployment.[22] Offensive

force modernization would be a proper immediate response to evidence of Soviet BMD breakout.

Also destabilizing might be a condition wherein both sides possess strategic defenses which would be effective against limited attacks, such as retaliatory strikes by a force that had been degraded by pre-emption. During this later phase of a defense transition it would be particularly important for stability that U.S. offensive forces retain a robust deterrent threat. Indeed, it would be critical to maintain that deterrent threat until U.S. defensive forces have achieved a very high level of effectiveness against Soviet preemptory and retaliatory strikes. Consequently, even during later phases of a defensive transition, when the U.S. may have deployed several layers of a planned multi-layered defensive system, to neglect offensive force modernization in the face of Soviet defensive advances would be perilous.

In short, the role of offensive forces during any defensive transition will be essential to the stability of that transition, whether it ultimately provides a "total" defense capability or a more limited protection of selected U.S. strategic assets. Maintaining and modernizing U.S. strategic offensive forces is essential to stability during any defensive transition. The question of, "how can we get from here to there safely?" is answered by maintaining offensive strategic retaliatory forces during the transition process.

It is possible that no set of defensive systems will ever provide a comprehensive defense against the Soviet Union. If it turns out to be true that the capability to defend urban/ industrial assets effectively will remain beyond reach,[23] then the deterrent role of offensive forces will remain unchanged; deployed defensive systems would serve only to support an offensive deterrent, and perhaps to defend against limited or accidental attacks. However, if the SDI leads to a comprehensive defensive capability, what role might strategic offensive forces play following the transition?

In a "defense-dominated" environment offensive forces would play a reduced, but still important role. For example, offensive forces would still provide deterrence leverage in a defense dominant strategic environment. The Soviet leadership could never have absolute certainty that its defensive "astrodome" would function effectively. This residual doubt should provide some continuing offensive deterrent leverage. Offensive nuclear forces could still pose a threat to some important Soviet assets, particularly military forces, outside the defended zone. Additionally, offensive forces would help sustain stability in the event that the effectiveness of defensive systems proved temporary. Finally, U.S. offensive forces would retain utility, and perhaps an increasingly necessary utility, to deter third party provocations. In short, U.S. strategic offensive forces would retain an important deterrence role even following a transition to very effective strategic defenses.

A basic principle with which the U.S. must approach a defense transition is that stability requires that the U.S. be capable of posing a retaliatory threat to the highest of Soviet leadership values at least until it can provide a comprehensive defense for its own highest values. The role of offensive and defensive forces during and following a transition should be to ensure that the U.S. is never left in a dilemma where it possesses neither adequate defenses to protect its own society nor retains adequate offensive forces capable of putting at risk the Soviet leadership and its sources of military and political control. Offensive forces will play an extremely important, indeed vital role in preserving a condition of stability during any defensive transition. If the transition does not lead to a comprehensive capability to defend U.S. urban/industrial assets, then offensive forces will retain the lead role in preserving stability.

If the defense transition does ultimately achieve a comprehensive defense capability, then the role of offensive strategic forces in maintaining stability will be reduced significantly but certainly not by the extent that some suggest. For example, Secretary of Defense Weinberger recently stated concerning strategic defense, "If it works, we will be able, I hope to eliminate the need to keep, maintain, and continue to modernize offensive weapons. That would be particularly true if the Soviets develop comparable technology, and I assume they will."[24] This, unfortunately, is unrealistic. Deterrence, even in a "defense dominant" strategic relationship, cannot be predicated solely upon a defensive shield. The achievement of a very effective defensive capability, an "astrodome," would not permit the U.S. to eschew a strategic offensive "sword." The deterrence functions required by U.S. commitments cannot stem only from even an immaculate defense; extended deterrence, for example, requires an offensive threat. What "defense dominance" could do would be to reorient the character of that threat away from now-traditional notions of punitive retaliation.

Following the Transition: Changing Requirements for Stability

Over the course of the last two decades stability has come to be closely identified with an offensive/ punitive approach to deterrence. Indeed, so close is the perceived connection between vulnerability and

stability that critics of the SDI charge that strategic
defense would endanger stability because it would
reduce perceptions of vulnerability. Until recently
official U.S. policy statements have reflected this
same position. For example, the Preamble of the 1972
ABM Treaty states that the parties were, "Considering
that effective measures to limit anti-ballistic missile
systems would be a substantial factor in curbing the
race in strategic offensive arms and would lead to a
decrease in the risk of outbreak of war involving
nuclear weapons."[25] Secretary of Defense Harold
Brown evidently opposed BMD at least in part on stability grounds, believing that "...the absence of a widely
deployed ABM system makes for stability."[26]

Nevertheless, it should be recalled that mutual
vulnerability is not a necessary condition for stability. Mutual vulnerability is by no means the only or
demonstrably most effective means of pursuing stability. Strategic defense and the reduction of U.S.
vulnerability to attack can also be compatible with
deterrence and the pursuit of stability. The critical
question that must be addressed is: How can strategic
defense support stability?

As mentioned above, stability and vulnerability
have become closely associated in American strategic
thought; this association is reflected in the punitive
approach to deterrence that has dominated U.S. policy
for almost two decades. However, deterrence certainly
functioned before the advent of nuclear weapons and
intercontinental means of delivery.[27] Deterrence
historically has been served by a capability to counter
the opponent's military establishment and by denying
the opponent any anticipation of being able to achieve
its objectives through the use of force. Such a
defensive-denial "deterrent" was always highly credible
because in exercising such a threat the state generally
was defending its highest values, not jeopardizing them
as would be the case if the U.S. exercised its current
deterrent threat. A transition to an extremely effective strategic defense would reintroduce a credible,
defensive-denial approach to deterrence. There are
several reasons to conclude that such a defensive
transition would support and perhaps even enhance
stability.

First, U.S. defenses could threaten to deny the
Soviet Union any plausible attainment of its "theory of
victory." Requirements of the Soviet theory of victory
include the following:[28]

-- preemptive counterforce strikes to destroy
opposing forces, and strategic defense to limit
damage to the Soviet Union;

-- the retention of strategic reserves for follow-on
   strikes;

-- total defeat of the opponent and all enemy
   forces;

-- the destruction of all centers of powerful opposition;

-- survival and recovery of the Soviet Union politically, economically, and militarily;

-- occupation of critical enemy territory, particularly Europe.

Effective strategic defenses would preserve U.S.
offensive forces and deny the Soviet Union the prospect
of defeating the U.S. and removing it as a powerful
center of opposition. The prospect of a protracted
stalemate at the strategic level should be highly
deterring to the Soviet Union given the Soviet prefer-
ence for a short, decisive conflict if central war
occurs. Soviet holdings and acquisitions would remain
at risk and vulnerable to a U.S. counter-offensive, and
a protected U.S. could have a military mobilization
potential that would foreclose any Soviet confidence
concerning the post-attack period. In contrast to the
current punitive orientation, a defensive deterrent
would be predicated upon the inability of the Soviet
Union to defeat the U.S., and the long-term threat that
U.S. military mobilization would pose to the Soviet
Union.
The U.S. military potential is indeed formidable
and should be an extremely effective instrument for
deterring attacks on the United States, its allies, and
its vital interests. The U.S. advantage in military
potential was a key to allied victory in both world
wars, and that fact undoubtedly is not lost upon Soviet
leaders. However, before the Soviet Union could be
made to respect U.S. mobilization potential, and be
deterred by it, that potential would need to be
effectively defended from Soviet attack. Effective
strategic defenses could establish a deterrent based on
"denial of victory," and a defensive deterrent should
be vastly more credible than the current punitive
approach. U.S. leaders will be perceived by opponents
as more willing to engage in defensive actions than
punitive, self-destructive actions.
For example, the credibility of the current U.S.
threat to engage in nuclear escalation on behalf of
NATO-Europe lacks credibility for some because of the
disastrous implications that the actual execution of
that threat would have for the United States. The

total vulnerability of the U.S. homeland surely would
weigh heavily against U.S. initiation of a process of
nuclear escalation. U.S. deterrence policy, in effect,
threatens to engage in actions that would likely be
self-destructive and could not be taken by an entirely
rational U.S. leadership. This situation is not conducive to deterrence credibility. Herman Kahn noted
the lack of credibility in a wholly offensive approach
to deterrence over twenty years ago:

> ... so far as the Soviets are concerned, the
> probability of [nuclear escalation] by us is
> small, particularly because we have made
> negligible preparations to ward off, survive,
> and recover from even a 'small' Soviet
> retaliatory strike. No matter how menacing
> they look, it will be irrational to attack
> and thus insure a Soviet retaliation unless
> we have made preparations to counter this
> retaliation.[29]

Kahn's comment is even more relevant today given the
vast increase in the Soviet capability to threaten the
American homeland. Soviet writers also have long
recognized this disconnect between the U.S. extended
nuclear commitment and actual U.S. interests.[30]
    Herman Kahn made an elementary, but extraordinari-
ly significant point concerning deterrence. A deterrent threat which could not be expected to be executed,
which would in fact be likely to lead to self-paralysis
by the side making the threat (in this case the U.S.),
is not likely to be a credible deterrent.
    An effective defensive capability should signifi-
cantly enhance the credibility of U.S. action because
the U.S. threat of nuclear escalation would not entail
potentially self-destructive consequences. A defended
America could more convincingly offer extended deterrence coverage to distant vital areas such as NATO-
Europe and the Persian Gulf.
    The U.S. and its NATO allies presently use the
deterring threat of nuclear escalation to compensate
for the potential inadequacy of local NATO forces to
deny the opponent victory in Europe. There have been
no fundamental changes in "extended deterrence" policy
since General Norstad remarked, "The function of shield
forces is really not to fight, not even to defend, but
to complete the deterrent."[31] More recently Senator Sam Nunn (D.-GA) observed that NATO's current
conventional posture constitutes "little more than a
delayed trip wire for early resort to nuclear escalation."[32] The reasoning behind such a policy is
that locally deployed forces need not be capable of
defending against an attack, but must be adequate to

make the opponent fear that a process of escalation
would begin following an attack. This fear of escalation rather than the capability actually to defend
against the attack, is expected to deter the opponent's
attack. Yet in the absence of strategic defense the
threat to initiate a process of nuclear escalation
could easily lead to national self-destruction. To
make such a threat on behalf of distant interests,
which if lost would not necessarily lead to national
self-destruction, cannot appear highly credible.

Effective defense of the U.S. homeland should
contribute to a Soviet belief that the U.S. would
abide by its commitment to NATO. In short, effective
strategic defenses should make U.S. commitments to
allies more believable and thereby enhance extended
deterrence.

Another factor which should contribute to stability after the defensive transition will be the continuing vulnerability of Soviet forces and assets in
regions outside the coverage of Soviet strategic
defenses. A Soviet offensive in Europe would still
entail the risk of a U.S. horizontal escalation,
including U.S. retaliation in another theater of war.
If the U.S. homeland were fully defended, the Soviet
leadership would be much more likely to believe that
the U.S. would engage in such escalation. Consequently
strategic offensive nuclear forces capable of theater
employment would retain an important deterrence role
following the transition.

Strategic defenses would not make the strategic
environment "safe" for nuclear or non-nuclear provocations. Such provocations would still entail severe
risks. Indeed, the Soviet leadership could perceive
these risks to be more deterring than the current condition of mutual vulnerability because U.S. leaders
will be perceived a more ready to act in the context of
effective defenses.

It is impossible to state with complete confidence
that a comprehensive defense would provide a more
effective deterrent than exists under the current punitive approach to deterrence. However, it is clear that
a "defense dominant" strategic relationship would contribute to the continuation of deterrence when:

--   The U.S. possessed a defensive capability to deny
      the Soviet Union its "theory of victory" in
      central war;

--   The Soviet leadership feared the domestic
      political repercussions of engaging the U.S. in a
      long and stalemated central war;

-- The U.S. posed a military mobilization potential that could leave no room for Soviet confidence following a theater or central war;

-- The Soviet warplanners and leaders could not be absolutely certain that their strategic defenses -- when put to the test -- might not fail spectacularly and catastrophically;

-- The Soviet leaders were aware of the continued vulnerability of Soviet land and naval forces outside their defended zones to the threat of U.S. nuclear escalation.

Defense Dominance and Instability

The factors contributing to a defensive-oriented stability are not the only implications of "defense dominance." There are several factors that could also have destabilizing effects. For example, effective defenses against intercontinental means of nuclear attack could lead the Soviet Union to concentrate on special operations and sabotage -- perhaps involving nuclear, chemical, and/or biological weapons.

The open nature of U.S. society, thousands of miles of open borders, and the potential vulnerability of the U.S. NCA and C^3I to limited attacks,[33] could result in Soviet exploitation of special operation threats. If so, the Soviet Union would have an advantage in offsetting strategic defenses through special operations, and a particularly potent incentive to exploit this avenue if U.S. offensive and defensive capabilities were very effective.

Second, effective Soviet strategic defenses would greatly reduce U.S. targeting access to the Soviet citadel -- especially against any limited strategic targeting. This could degrade or negate U.S. limited retalitory threats supporting extended deterrent coverage for NATO-Europe and the Persian Gulf. Limited nuclear threats backstopped by the ultimate sanction of "catastrophic retaliation" are the key to the U.S. theory of extended deterrence and escalation control. Consequently, a well-defended Soviet Union might consider itself more at liberty to exploit its military advantages in and around Eurasia because the Soviet Union might be able to parry the threatened limited attack. Intermediate steps on the continuum of U.S. escalation threats (theater nuclear, and limited strike threats against the Soviet zone of the interior) might no longer be persuasive. Similarly, the "ultimate" threat of large-scale strategic nuclear use against the Soviet Union would be less deterring. Consequently, a transition to effective strategic defenses might undercut stability in vital distant areas.

Finally, an effective Soviet defensive capability would nullify whatever deterrent leverage exists as a result of independent British and French nuclear forces. Although it is impossible to "prove" what British and French forces contribute to deterrence of war in Europe, its seems likely that Soviet inhibitions are higher because of independent French and British strategic forces. A defensive transition would relieve the Soviet Union of whatever contribution to deterrence these independent forces now provide. Of course, this is a basic reason why some Europeans have expressed concern regarding the SDI.[34]

This point underscores the difference between a defended approach to extended deterrence, and the current punitive approach. The U.S. threat of "catastrophic retaliation" would no longer provide the ultimate sanction of a "flexible response" doctrine, although the U.S. commitment to NATO would still be made manifest by the local stationing of U.S. forces. Rather, extended deterrence would be based upon a multi-layered threat: (1) the possibility that a European conflict would escalate to a global conflict wherein Soviet forces outside of their defended zone would be vulnerable to attack (horizontal escalation); (2) NATO escalation to the use of tactical nuclear weapons to help deny a local victory to the Soviet Union; and (3) ultimately, a mobilized U.S. would contest Soviet control of occupied territory. These sanctions might appear less fearsome to the Soviet Union than a threat of catastrophic retaliation, but should be much more credible than the current situation where the United States risks annihilation through any nuclear escalation in Europe and yet relies upon such a threat to deter Soviet and Warsaw Pact invasion.

The effectiveness of NATO's local stopping power would be extremely important to a victory-denial deterrent in Europe. The fruit of the U.S. industrial mobilization potential would take some time to come into play in such a conflict, and if Europe had been defeated and occupied during that period the U.S. could have the formidable task of attempting to repeat its World War II performance in an era where a Normandy-style beachhead could be eliminated by a few tactical nuclear weapons of relatively low yield. Consequently, NATO's capability to preserve enough time and space for the U.S. to respond would be crucial. That problem is, of course, little different from current NATO requirements to delay and survive Soviet attack. The difference being that in a defense dominant strategic relationship the U.S. is more likely to be able and willing to react swiftly and prepare a long-term counter offensive, if necessary. This difference should enhance deterrence.

Again, it is impossible to know with any certainty whether deterrence is more effective when predicated on a punitive offensive threat or a defensive-denial threat. The former would have more immediately severe consequences while the latter would be much more credible. The point is that a defensive stalemate should not liberate Soviet aggressive designs any more than does the current condition of U.S. vulnerability which could paralyze U.S. decision-makers and prevent them from taking action during any escalating clash of arms with the Soviet Union.

It is far from clear that strategic defenses would leave Eurasia more vulnerable to Soviet aggression than does the current condition. What is clear is that a defensive transition would realign U.S. interests with U.S. extended deterrence commitments.

A transition to very effective strategic defenses should enhance the credibility of the U.S. alliance commitment and thereby contribute to stability. Whether enhancing credibility compensates, or more than compensates, for the likely reduction in the immediacy and severity of the threat (because of Soviet defenses), is an important question which cannot be answered with any certainty. What can be claimed is that highly effective defenses need not be inconsistent with deterrence. Defenses simply would supply a different disincentive to Soviet aggression than the threat of immediate nuclear retaliation against the Soviet Union. In addition, if deterrence should fail strategic defenses would provide a margin of safety for the U.S. that can not exist under an offensive-punitive approach to deterrence.

To maximize the prospects for stability following a transition to very effective defenses the U.S. must maintain modernized strategic offensive forces. Offensive forces would continue to contribute to stability in several ways.

First, continued modernization of U.S. strategic offensive forces should help convince Soviet leaders that an offensive and defensive advantage, i.e., strategic superiority, could never be achieved and should not be pursued. Ironic as it may seem, an on-going U.S. offensive modernization program should actually help establish the basis for arms control negotiations. This is not to say that U.S. offensive and defensive programs will necessarily result in cooperative Soviet behavior. However, without both types of programs there would exist little possibility for successful negotiations.

Second, offensive forces will continue to provide some residual deterrent leverage. Neither side could have absolute confidence in its defenses as long as offensive forces are maintained, and this uncertainty would be important for post-transition stability. This offensive deterrent "hedge" would also be extremely important as a precaution against the possibility that unexpected technical developments could compromise the effectiveness of the defense. Offensive forces would retain the critical role of providing a deterrent backstop for defensive forces that might come under serious technical challenge.

Third, both superpowers will have reason to maintain an offensive, coercive capability against third powers. The continuing development of PRC nuclear forces alone will ensure that the Soviet Union will not relinquish an offensive nuclear capability. The Chinese currently have limited strategic nuclear capabilities (the T-5 ICBM, experimental T-4 ICBM, and CSS-NX-14 SLBM exist in very limited numbers),[35] but during the next 20-30 years when the defensive transition takes place, Chinese forces will improve quantitatively and qualitatively. During this period nuclear weapons may be developed in many other states, some of them unfriendly to the U.S., giving additional reason for both the U.S. defensive transition and the maintenance of an offensive, coercive capability.

Finally, offensive nuclear capabilities would have to be sustained to provide the threat of nuclear escalation associated with the capability for nuclear land and naval warfare outside the Soviet defense zone. This continued threat of nuclear escalation would help stabilize the post-transition period. In particular, it would be essential for constraining Soviet willingness to exploit its military advantages in and around Eurasia.

Offensive Forces in a World of Limited Defenses

It may well prove impossible to provide the defensive capability assumed above, i.e., comprehensive protection for urban/industrial (U/I) assets. The requirements for such a capability are formidable and perhaps beyond reach.[36] For example, assuming a (not unrealistic) future Soviet threat of 20,000 to 30,000 weapons, even overall defense effectiveness of 99% would permit leakage of 200-300 nuclear weapons. If concentrated on American urban/industrial areas this number of "leakers" clearly could cause "unacceptable damage."

If the U.S. and Soviet Union must anticipate several hundred or more "leakers," what would be the effect on the U.S.-Soviet deterrence relationship? The difference in U.S. and Soviet value hierarchies is significant in this regard. There is little doubt that urban/industrial assets constitute the highest values of the U.S. In contrast, U.S. deterrence policy is

founded on the belief that the highest values of the
Soviet leadership consist of political and military
power, and the instruments for securing that
power.[37] This asymmetry in value structures could
be extremely disadvantageous for the U.S. because,
while several hundred penetrating warheads could
threaten the highest values of the U.S., the Soviet
Union has gone to extreme lengths to protect its
control structure from nuclear attack. As two of the
foremost American "Sovietologists" observe:

> There is no certainty that any group govern-
> ing a large, heterogeneous population can
> design controls that would be effective
> throughout a nuclear war and in its after-
> math. If such a system could be perfected,
> it probably exists in the Soviet Union....
> This requirement is a basic tenet of Soviet
> military doctrine.[38]

Former Secretary of Defense Harold Brown has
estimated that 1,000-2,000 penetrating hard-target
capable warheads would be the minimum required to pose
a threat to the superhardened Soviet control structure.[39] It is absolutely critical to note that a
less-than-perfect defense structure that allows several
hundred penetrating warheads might provide adequate
protection to the highest values of the Soviet leadership, yet leave the highest of U.S. values at risk.

Such asymmetry in vulnerability could have highly
destabilizing consequences in an acute crisis (precisely when the strategic deterrent would matter most).
The Soviet leadership could (rightly) consider itself
at much greater liberty to press its position and
escalate a crisis than would U.S. leaders. Soviet
highest values would be less at risk than would U.S.
values and consequently the Soviet leadership could
confront the U.S. with confidence. Strategic defense
in this case would provide the Soviet Union with a form
of strategic superiority and escalation dominance. In
this case a limited strategic defense could preclude a
U.S. defensive "victory-denial" deterrent and leave the
U.S. vulnerable to defeat.

Such a condition would cripple the American
extended deterrent. The Soviet leadership could pursue
highly provocative actions in Europe, the Persian Gulf
or elsewhere with the knowledge that the U.S. would be
extremely unlikely to initiate a nuclear escalation
process, or if the U.S. did escalate the conflict, the
Soviet Union could emerge the victor. Such a situation
is not appropriate for extending deterrent coverage to
distant allies and friends.

These considerations of a "leaky" strategic defense indicate that a defensive transition could lead to a most destabilizing condition: defenses inadequate to preserve the highest U.S. values and a retaliatory force incapable of threatening the opponent's highest values. Stability requires the maintenance of a U.S. offensive threat that is at least <u>functionally</u> equivalent to the Soviet threat, i.e., U.S. retention of a threat against the highest Soviet values -- a threat which would require a statistically superior penetration capability for U.S. offensive forces. If the Soviet Union could hold U.S. urban/industrial areas at risk, the U.S. must retain the ability to hold Soviet leaders and their military and political power at risk. The asymmetry in U.S. and Soviet values would have to be offset by an asymmetry in offensive capabilities -- with the advantage held by the United States.

In this case, a less than comprehensive U.S. defensive capability would not permit the U.S. to pursue a less dynamic offensive weapons modernization program. Indeed, because U.S. offensive forces would have to penetrate Soviet defenses in substantial numbers and at selected (probably defended) locations, the U.S. strategic offensive modernization program could perhaps require augmentation.

In short, the "most realistic" defensive transition scenario, one acknowledging a degree of leakage, would require the maintenance of an extremely capable U.S. offensive capability. Given the unique hard-target lethality and control available to ICBMs, and the enhanced survivability that could be provided for hardened silos by even modest active and passive defenses, such a defensive transition would entail a continuing, and probably even increased role for ICBMs in holding the highest values of the Soviet leadership at risk for the purpose of maintaining stability.

Summary and Conclusion

The U.S. is embarking upon a strategic defense initiative. Presently, the SDI is intended only to examine the feasibility and policy implications of strategic defense -- to provide an informed basis for a later decision concerning possible deployment. If a decision for deployment is made, the SDI objective chosen is extremely important. Does the U.S. opt for a limited defense for retaliatory forces or a comprehensive defense for cities? The answer has important implications for offensive force modernization -- including issues of basing modes, funding concerns, design requirements, and the fundamental question of what role offensive forces play during and after a defensive transition.

Very effective comprehensive defenses should serve to discourage any Soviet inclination to combat the U.S. directly, and would make U.S. participation in theater conflicts more credible. Therefore, the Soviet leadership should be deterred from both central and theater conflicts with the U.S. and it allies. However, the transition itself could be destabilizing unless the U.S. maintains an effective offensive deterrent during that period. Strategic offensive force modernization would continue to play an important deterrent role even if seemingly perfect defenses were acquired, but would be absolutely essential to stability if the United States had less-than-perfect defensive capabilities.

There are at least four possible futures to consider when assessing the implications of strategic defenses for deterrence:

(1) Area Defense - Effective Offense: a capability to defend effectively both U.S. U/I assets and the U.S. retaliatory forces necessary to threaten the highest Soviet values.

(2) Area Defense - Theater Offense Only: a U.S. capability to defend U/I assets effectively, and strategic forces suitable only for targeting Soviet assets largely outside the Soviet Union.

(3) Point Defense Only - Effective Offense: a U.S. defensive capability inadequate to provide comprehensive coverage for U/I assets, but sufficient to protect a retaliatory force capable of threatening the opponent's highest values.

(4) Inadequate Defense - Inadequate Offense: U.S. defenses inadequate to preserve either U/I assets or a retaliatory force capable of threatening the opponent's highest values.

If U.S. defenses are incapable of comprehensive coverage, and the U.S. does not maintain an effective offensive threat, then the most destabilizing condition (number four) will be established. If U.S. defenses are highly effective, providing comprehensive coverage for cities, then the offensive deterrent forces will retain an important but less critical role. However, in the more likely event that a comprehensive defense for urban/industrial assets remains uncertain for the foreseeable future, offensive retaliatory forces, and the ICBM in particular, will provide essential support for deterrence stability.

NOTES

1. For the full text of the President's March 23rd speech see, The New York Times, March 24, 1983, p. 20.

2. Quoted in Aerospace Daily, March 29, 1983, pp. 161-162; and Charles W. Cordry, "Weinberger Says Total Defense Is Sought For U.S.," Baltimore Sun, March 28, 1983, p. 1.

3. See the discussion of these studies in, Clarence Robinson, "Panel Urges Defense Technology Advances," Aviation Week and Space Technology (October 17, 1983), pp. 16-18.

4. Quoted in, "Keyworth: Space-Based Defense Possible," Air Force Times, Oct. 31, 1983, p. 29.

5. Quoted in Fred Hiatt, "Limited ABM is Urged To Protect U.S. Missiles," The Washington Post, March 6, 1984, p. A-30.

6. President's Commission on Strategic Forces, Report of the President's Commission on Strategic Forces (April 6, 1983), p. 6.

7. This letter is discussed in, "Washington Roundup," Aviation Week and Space Technology, Vol. 120, No. 3 (Jan. 16, 1984), p. 13.

8. See recent Congressional testimony on this matter by Dr. Robert Cooper, Director of DARPA; reported in, Walter Andrews, "Pentagon Aide says Soviets Deploying Weapon Convertible to ABM Defense," Washington Times, March 9, 1984, p. 3.

9. Quoted in Defense Daily, March 9, 1984, p. 50.

10. See for example, William Davis, "Ballistic Missile Defense Will Work," National Defense, Vol. 66, No. 373 (December 1981), p. 16.

11. See Hiatt, op. cit., p. A-30.

12. Quoted in, Soviet Aerospace, February 27, 1984, p. 57.

13. For analyses of passive defense for U.S. silos see, National Institute for Public Policy, U.S. National Security Policy Implications of Superhard Silos (Fairfax, VA: NIPP, March 1984).

14. The CIA reportedly has concluded that there appears to be some evidence to support such an idea. See, Clarence Robinson, "Soviets Accelerate Missile Defense Efforts," Aviation Week and Space Technology, Vol. 120, No. 3 (Jan. 16, 1984), pp. 14-16.

15. James Woolsey made this point at a Washington, D.C. conference, introducing Nuclear Arms: Ethics, Strategy, Politics (San Francisco: ICS Press, 1984), edited by James Woolsey, Mayflower Hotel, April 11, 1984.

16. This notion of a preemptive Soviet attack on U.S. space-based BMD deployment often is suggested as one reason that an SDI incorporating space-based weapon components would be dangerous. See for example, Edgar Ulsamer, "The Long Leap Toward Space Laser Weapons," Air Force Magazine, Vol. 64, No. 8 (August 1981), p. 63.

17. Department of Defense, Soviet Military Power (Washington, DC: USGPO, April 1984), p. 34.

18. Discussions of "crisis instability" are rampant in U.S. strategic theory. See for example, Thomas C. Shelling, Arms and Influence (New Haven, Conn.: Yale University Press, 1966), p. 225; and Jerome Kahan, Security in the Nuclear Age: Developing U.S. Strategic Arms Policy (Washington, DC: Brookings Institute, 1975), pp. 272-273, 282. More recently see, Strobe Talbott, "The Case Against Star Wars Weapons," Time (May 7, 1984), pp. 81-82.

19. Harold Brown, Department of Defense Annual Report Fiscal Year 1980 (Washington, D.C.: USGPO, January 25, 1979), p. 61.

20. See the discussion in, Rebecca V. Strode, "Space-Based Lasers For Ballistic Missile Defense: Soviet Policy Options," in Keith Payne, ed., Laser Weapons In Space: Policy and Doctrine (Boulder, Colorado: Westview Press, 1983), pp. 107-125.

21. See John Erickson, "The Soviet View of Deterrence: A General Survey," _Survival_, Vol. 24, No. 6 (November/December 1982), p. 246.

22. _National Security Issues, 1981 Symposium_ Mitre Corporation, Bedford, Mass., October 13, 1981.

23. See for example, Carl Sagan, "Nuclear War and Climate Catastrophe: Some Policy Implications," _Foreign Affairs_, Vol. 67, No. 2 (Winter 1983/84), p. 282; and, Wolfgang Panofsky, "The Mutual-Hostage Relationship Between America and Russia," _Foreign Affairs_, Vol. 52, No. 1 (October 1973), pp. 109-118.

24. Quoted in an interview with _Omni_ magazine, September 1983, p. 114.

25. U.S. Arms Control and Disarmament Agency, _Arms Control and Disarmament Agreements_, Publication 105 (Washington, D.C.: USGPO, August 1980), p. 139.

26. "Brown Says An Accord Won't Make Teheran Into A 'Buddy' of U.S.," _New York Times_, January 19, 1981, p. A-12.

27. See a brief discussion on ancient approaches to deterrence in, Thomas Schelling, "Surprise Attack and Disarmament," in _NATO and American Security_, Klauss Knorr, ed. (Princeton, N.J.: Princeton University Press, 1959), pp. 188-189.

28. For an excellent summary discussion of the Soviet theory of victory see John J. Dziak, _Soviet Perceptions of Military Power: The Interaction of Theory and Practice_ (New York: Crane, Russak & Company, 1981), pp. 27-28.

29. Herman Kahn, _On Thermonuclear War_ (Princeton, N.J.: Princeton University Press, 1961), pp. 132-133.

30. See for example, V.D. Sokolovskiy, _Soviet Military Strategy_, 3rd Edition, Harriet Fast Scott, ed. (New York: Crane Russak, 1975), pp. 55-57.

31. Quoted in Glenn Snyder, Deterrence and Defense (Princeton: Princeton University Press, 1961), p. 127.

32. Quoted in Helen Dewar, "Reagan Fights Plan to Remove 90,000 Troops From Europe," Washington Post, June 20, 1984, p. 1.

33. For a discussion of the vulnerability of U.S. C^3 facilities see, Paul Bracken, The Command and Control of Nuclear Forces (New Haven, Conn: Yale University Press, 1983), and Desmond Ball, "Can Nuclear War Controlled?," Adelphi Papers, No. 169 (London: IISS, Autumn 1981). On the great potential difficulty of identifying and finding the Soviet leadership see, Samuel T. Cohen and Joseph D. Douglass, Jr., "Selective Targeting and Soviet Deception," Armed Forces Journal (September 1983), pp. 95-101.

34. See for example, Hans-Heinrich Weise, "Amerikanische Pläne für ein weltraumgestütztes Raketen-Abwehrsystem," Europa Archiv, Folge 13 (Juli 10, 1984), p. 406

35. See The Military Balance 1983-1984 (London: IISS, Autumn 1983), p. 84.

36. See, Directed Energy Missile Defense in Space -- A Background Paper (Washington, D.C.: U.S. Congress, Office of Technology Assessment, OTA-BP-ISC-26, April 1984), p. 81.

37. See for example, President's Commission on Strategic Forces, Report of the President's Commission on Strategic Forces (April 6, 1983), p. 6.

38. Harriet Scott and William Scott, The Soviet Control Structure: Capabilities For Wartime Survival (New York: Crane Russak, 1984), p. 129.

39. See Harold Brown, <u>Thinking About National Security</u> (Boulder, CO: Westview Press, 1983), pp. 74-75.

6
Two New U.S. ICBMs–
Better than One?

Michael Ennis

The President's Commission on Strategic Forces, the Scowcroft Commission, set in motion some important changes in the U.S. ICBM modernization program. The Commission reaffirmed the critical importance of a modernized land-based missile force, observing that "ICBMs are especially effective in deterring Soviet threats of massive conventional or limited nuclear attacks, because they could most credibly respond promptly and controllably against specific military targets and thereby promptly disrupt an attack on us or our allies."[1]

The Commission fully supported deployment of Peacekeeper ICBMs, but sole reliance upon the Peacekeeper as the single ICBM of the future was rejected. Given the threat of increasingly numerous and accurate Soviet ICBMs, the Commission concluded that prudence dictated ICBM diversity.

> ...the preferred approach for modernizing our ICBM force seems to have three components: initiating engineering design of a single-warhead small ICBM, to reduce target value and permit flexibility in basing for better long-term survivability; seeking arms control agreements designed to enhance strategic stability; and deploying MX missiles in existing silos now to satisfy the immediate needs of our ICBM force and to aid in that transition.[2]

President Reagan endorsed the findings of the Scowcroft Commission and is using their report as the blueprint for U.S. strategic modernization. The President's program now includes the development of a small missile, a hardened mobile launcher and follow-on basing technology; a shift in the U.S. negotiating position at the START talks; and, the production of Peacekeeper and deployment in existing Minuteman silos.

By introducing the small ICBM as a follow-on system the Scowcroft Commission sought to bypass troublesome strategic and survivability questions, issues which seemed to defy a politically acceptable solution. Ironically, it is in part the current wave of enthusiasm for the SICBM that has fueled opposition to the Peacekeeper. Opponents of the Peacekeeper have argued that the Scowcroft Commission recommendation for the small ICBM remains valid even if the Peacekeeper is cancelled. The small ICBM, they contend, provides a sufficient guarantee of deterrence. Columnist Tom Wicker, for example, insisted that "1,000 Midgetmen could be deployed as soon as and for less money than 100 MXs and would provide a more stable deterrent."[3] A substantial number of Senators and Representatives appear to share this sentiment for the small ICBM. As Senator Moynihan has argued:

> Small, mobile missiles are the way of the future in intercontinental ballistic missiles, and also the way to stability in the nuclear balance, for they are infinitely better suited to deterrence than the behemoths of which the Soviets have hitherto seemed so fond. Small mobile missiles are also better suited to deterrence than MX...I will offer today an amendment that, while halting the MX deployment, would double the money spent on research and development for a small, single-warhead ICBM.[4]

Three basic alternatives for ICBM modernization are examined here -- deployment of the Peacekeeper only, deployment of the small ICBM only, and tandem deployment of the Peacekeeper and small ICBM. These alternatives are then examined from the standpoint of ideal military utility and ideal arms control characteristics. The specific questions to be raised include:

1) What is the unique function that ICBMs can play in U.S. strategy and policy?

2) What are the attributes of an ideal ICBM force?

3) How does the deployment of 100 Peacekeepers in Minuteman silos compare to the ideal ICBM force?

4) How does the SICBM deployment compare to the ideal ICBM force?

5) How does the tandem deployment of both Peacekeeper and SICBM, as proposed by the Scowcroft
Commission, compare to the ideal ICBM force?

6) What ICBM force best suits the multiple
requirements of U.S. strategy and policy:

- Peacekeeper alone?

- SICBM alone?

- Peacekeeper/SICBM together?

Attributes of an Ideal ICBM Force

The controversy over strategic modernization, at
its core, is the search for a perfectly tailored ICBM
force. Such an ideal force would satisfy all the
deterrence and arms control requirements without contributing to strategic instability. This ideal force
would have the following characteristics:

(1) Reduces Risk of War: A force substantial
enough to deter without posing a first strike
threat.

(2) Survivability: The ability to survive an
enemy first strike and retaliate in a
controlled manner over an extended time.

(3) Accuracy: CEP sufficient to destroy hardened
targets with high confidence.

(4) Prompt and Flexible Response: The ability to
retaliate quickly and retarget as necessary
in a short time across a broad range of
targets.

(5) Enduring Communications: Reliable C^3I
links to the NCA to insure uninterrupted and
effective command and control.

(6) System Reliability: Efficient performance
under wartime conditions.

(7) Responsiveness to Countermeasures: The ability to keep one step ahead of enemy weapons
developments and strategic innovations.

(8) Crisis Stabilizer: The system should present
no first strike threat to an enemy, and

should also be relatively immune to a first
strike by an enemy force.

(9) <u>Symbol of National Will</u>: The system should
be a clear demonstration of a national willingness to keep pace with the military
programs of a major adversary.

(10) <u>Timeliness of Deployment</u>: The system should
be fielded in a time to deter near-term as
well as long-term threats to national security.

(11) <u>Can Penetrate Enemy Defenses</u>: Has the capability of penetrating the active and passive
defenses of key targeted points of potential
adversaries.

(12) <u>Arms Race Stability and Cost Control</u>: The
system should not provoke the adversary to
assume a more threatening posture or to
initiate countermeasures that force ultra-
expensive responses by the U.S.

(13) <u>Arms Negotiation Leverage</u>: The system should
carry a sufficiently potent threat to induce
the adversary to negotiate on force reductions.

(14) <u>Conformity With Present Arms Control Agree-
ments</u>: The system should not violate arms
control agreements already in effect.

(15) <u>Verifiability</u>: The force should be deployed
in modes that are adequately and effectively
verifiable by the national technical means of
the other side.

Peacekeeper Deployment in Silos

Until accuracy and MIRV improvements in Soviet
missiles theoretically made them vulnerable, U.S. ICBMs
possessed almost all the desirable qualities of an
ideal deterrent force.

A common agreement that the ICBM force should be
modernized has been overshadowed by disagreement over
how this should take place. Some have concluded that
new strategic modernization should be attempted only if
it restores to ICBMs its "ideal deterrent" status. As
the Scowcroft Commission noted, "[the past performance
of ICBMs] has led many to continue to replicate those
decades of Minuteman history, and in doing so try not
only to meet these objectives, but to do so in a single

way of basing a single type of ICBM that would have all of these desirable characteristics."[5]

Judged against the ideal system the Peacekeeper has many of the qualities desired in an ICBM. Their high reliability, accuracy, rapid retargetability, prompt delivery, and secure links to the NCA insure that the Peacekeeper can target what Soviet leaders value most -- their political and military control mechanisms and their own lives. If the Soviets are unable to neutralize this capability, they would have little incentive to initiate an attack. The proposed deployment of only 100 missiles will fall short of a disarming first strike capability, but it will deliver to Soviet warplanners the message that they cannot expect to prevail in a conflict.

In addition, the Peacekeeper complies with the SALT I and SALT II provisions governing the deployment of ICBMs. The proposed deployment in silos permits ready verification of existing arms control agreements, as well as any treaty likely to be negotiated in the future. Also, the Peacekeeper can be deployed rapidly. The missile has already entered production and has completed a number of highly successful test flights. The 1986 IOC remains viable.

On the other hand, in several important categories the Peacekeeper falls short of the ideal. In particular, no politically acceptable basing mode could be found that assured the system's survivability beyond the first exchange. Regardless of the validity of the "ICBM vulnerability does not translate into triad vulnerability" argument, skeptics continue to view large, stationary ICBMs as unprotectable invitations to a Soviet first strike.

The Peacekeeper, as it will be initially based, lacks the desired measure of enduring survivability. The large warhead-to-missile ratio insures that they would be selected as an early target for Soviet warplanners. Peacekeeper's ability to deliver a controlled, measured response over an extended period of time may therefore be limited. Thus, the Peacekeeper force can make a valuable contribution to U.S. strategic posture, but lacks some of the characteristics of an ideal ICBM force. (See Table 6.1.)

Attempts to transform the Peacekeeper into the single, perfect weapon have only fueled opposition. It was the conclusion of the Scowcroft Commission that all the desired characteristics could not and need not be incorporated into a single weapon. "By trying to solve all ICBM tasks with a single weapon and a single basing mode in the face of the trends in technology," they argued, "we have made the problem of modernizing the ICBM force so complex as to be virtually insoluble."[6]

TABLE 6.1
Rating Peacekeeper Against the Ideal
ICBM Force

Criteria	Peacekeeper* Force	Ideal ICBM Force
1. Reduces Risk of War	Good	Excellent
2. Enduring Survivability	Poor to Good	Excellent
3. Accuracy	Excellent	Excellent
4. Reliability	Excellent	Excellent
5. Enduring C^3I	Fair	Excellent
6. Prompt/Flexible Response	Excellent	Excellent
7. Responsiveness to Countermeasures	Fair	Excellent
8. Crisis Stabilizer	Fair	Excellent
9. Symbol of National Will	Excellent	Excellent
10. Timeliness of Deployment	Excellent	Excellent
11. Capability vs. Enemy BMD	Excellent	Excellent
12. Arms Race Stability and Cost Control	Good	Excellent
13. Arms Negotiation Leverage	Excellent	Excellent
14. Conformity with Present Agreements	Excellent	Excellent
15. Verifiability	Excellent	Excellent

* Assumes current plans for deployment of 100 Peace-keeper ICBMs in present silos.

Small Mobiles Versus a Mixed ICBM Force

Not everyone in Congress agreed with the Scowcroft Commission that one new kind of ICBM could not satisfy all U.S. ICBM requirements.[7] Some want to drop Peacekeeper and substitute small mobile ICBMs instead.

The real debate at present is between those advocating a dual deployment of Peacekeeper (IOC 1986) and small ICBMs (1992) and those advocating a crash program for SICBMs to be deployed alone.

The argument in favor of deploying only the small missile rests on the questionable assumption that SICBM comes closer than Peacekeeper to satisfying the requirements of the ideal ICBM force, particularly with regard to survivability and arms control. However, the case for dual deployment does not rest on the fact that one missile system is better than the other. It is simply that together they form a deterrent force which is superior to either system by itself. The "SICBM only" option must be judged, not against the Peacekeeper, but rather against the Peacekeeper/SICBM tandem. By measuring both options against each of the characteristics of the ideal ICBM force, a more complete comparison is drawn. A detailed comparison of the two U.S. ICBM deployment options is discussed attribute by attribute on the pages that follow.

Reducing Risks of War: Peacekeeper/SICBMs or SICBMs Only?

Some have made the argument that deployment of high-accuracy ICBMs like Peacekeeper or the small ICBMs is destabilizing because they would put decisionmakers in a "use or lose" crisis dilemma. This, the argument goes, puts a hair trigger on nuclear war and opens both sides to the risk of accidental war or one due to miscalculation. What this viewpoint misses is that there is a triad of strategic forces on both sides. ICBM vulnerability is not triad vulnerability. Therefore there is no hair trigger put on nuclear war due to new ICBM deployments. The "counterforce weapon is dangerous" theory is valid only in the special case where all three elements of the triad of forces are vulnerable simultaneously. This is not the case today, nor is it likely in the next decade.

The tandem of Peacekeeper and small ICBMs could provide the United States with the same hard target coverage a force composed only of small ICBMs with an equivalent number of survivable warheads. However, the Peacekeeper component of the tandem deployment option can be fielded six years earlier than the small ICBM, and therefore can provide the target coverage necessary to implement the U.S. countervailing theory of deterrence that much earlier. This could be invaluable to the deterrence of Soviet-initiated aggression during any intense U.S.-Soviet crisis between 1986 and 1992. If the U.S. is serious about implementing its preferred deterrence strategy there is no alternative to deploying the Peacekeeper along with the small ICBMs.

Survivability: Peacekeeper/SICBMs or SICBMs Only?

The attractiveness of the small ICBM lies in the improved survivability which mobility might offer. A fully self-contained missile mounted on a hardened mobile launcher would present no fixed site which Soviets could target. Current plans envision the small missile operating over territory on government reservations. In a crisis the missiles could be dispersed and moved at random around the reservations. With no sure way to predict the precise location at any given moment, to destroy such a force the Soviets would probably have to barrage the SICBM's entire operating area. The effectiveness of such a barrage would depend on the hardness of the vehicle, the size of the operating area, and the number of Soviet weapons used in the attack.

Using such a large part of its resources against the small missile would limit the effectiveness of any Soviet first strike. With the wide areas available to the SICBM, an attack could require a disproportionate number of warheads directed against the mobile U.S. missiles. Further, there would be no certainty that all SICBMs were operating in the areas targeted. Simply moving outside of its normal operating area could greatly increase the small missile's chances of survival.

The difficulties in attacking a mobile system should help deter a Soviet strike, but the deployment of the Peacekeepers and the small missiles together would appear to be more survivable than SICBMs alone. A clear benefit of dual deployment is that the two missile combination complicates Soviet targeting in two different ways. The Peacekeeper, if it is eventually placed in superhard silos, might induce the Soviets to substitute fewer, larger yield warheads on their presently MIRVed ICBMs[8] and perhaps deploy on each a single, large, earth-penetrating warhead that would have more effect than a larger number of smaller yield warheads. For example, the Soviets may opt to replace the 10 warheads currently on the SS-18 with one very large yield weapon.

Conversely, targeting a small mobile missile creates the incentive to deploy a larger number of smaller yield warheads to barrage large operating areas. The best Soviet response to the SICBM may be to increase the number of warheads on the SS-18s and SS-19s. With further tests, it is thought that the Soviets could convert their 10-warhead SS-18s into missiles carrying perhaps 20 to 30 warheads.

The synergism between the Peacekeeper and the small missile creates a dilemma for the Soviet war-planner. The value of a first strike against silo-based ICBMs, even if it could be orchestrated success-fully, could be negated by the Soviet's inability to locate and strike the small mobile missiles. The Soviet Union could be forced to use a disproportionate amount of its strategic forces in barraging SICBMs, perhaps to the point where it lacked adequate resources to target remaining U.S. military forces and installa-tions. Maximizing the capabilities against one could minimize the capabilities against the other. This two-way stretch of strategic resources is particularly important because it would add to the uncertainty in the mind of Soviet warplanners.

The prospects for devising a survivable mobile basing scheme for the small missile are uncertain at present. Initial estimates envision launchers hardened to 100 psi or greater.[9] Nonetheless, such estimates remain theoretical. Numerous hurdles must be overcome before a truly survivable mobile ICBM comes off the assembly line. In the current euphoria over Midgetman it is often overlooked that mobile ICBMs have been studied for over two decades. The difficulty has always been devising a workable model.[10]

Accuracy: Peacekeeper/SICBMs or SICBMs Only?

Both the Peacekeeper and the SICBM will have greatly improved accuracy. The precision of the Peace-keeper's AIRS guidance system means that more hardened Soviet targets will be vulnerable. While the SICBM is still in the initial design stage, there is high con-fidence that its CEP will be in the Peacekeeper's class.[11] Therefore there may be no accuracy advan-tage enjoyed by one ICBM deployment option when com-pared to the other.

Prompt and Flexible Response: Peacekeeper/SICBMs or SICBMs Only?

Both the Peacekeeper and the small ICBM will be capable of prompt, controlled attacks against hardened military targets. Their C^3I is fully contained within the United States, thus simplifying the problem of two-way communications. Execution of an ICBM launch can begin within minutes of transmission of the NCA orders. Additionally both systems will have remote, rapid retargeting capabilities.

Enduring Communications: Peacekeeper/SICBMs or SICBMs Only?

The creation of survivable C^3I has been recog-nized as a matter of highest national priority.[12] The command and control of U.S. strategic forces resides with the National Command Authorities (i.e., the President and the Secretary of Defense and their successors). No potential attacker should be given reason to believe that the link between the NCA and U.S. ICBMs could be broken in a first strike.[13]

To assure communications with the SICBM, current C^3I programs would have to be adapted to mobility. The challenge to the SICBM designers is to adapt the C^3I technology designed for a fixed site system to a mobile system. While there are obstacles to be overcome, they are not viewed as insurmountable.[14]

It is true, however, that the nature of the Peacekeeper makes it less likely to lose C^3I capabilities. Because of its stationary position the communications links can be well shielded and carefully planned. And, since it is easier to defend a perimeter than protect a mobile vehicle, the Peacekeeper's C^3I can be better protected against domestic terrorism and Soviet special forces. The Peacekeeper/SICBM tandem provides a hedge against the unforeseen breakdown of either system and is more likely to provide enduring command, communications, and control than small ICBMs deployed by themselves alone.

System Reliability: Peacekeeper/SICBMs or SICBMs Only?

Silo-based ICBMs are the most reliable of strategic weapons. Redundancy of critial components and self-contained energy and environmental systems greatly reduce the maintenance requirements. It has been estimated that the Peacekeeper in silos will have a 95% Operational Readiness Rate.[15]

It is unlikely that a mobile SICBM could attain similar high readiness rates. The hardened mobile launcher and its accompanying support vehicles (C^3I, maintenance, fuel, security) will be subject to the jarring effects of off-road travel as well as a nuclear environment. The longer the time in the field, the greater the maintenance problems. If the SICBMs are dispersed for prolonged periods the reliability of the system will be in question. Therefore, the tandem deployment of Peacekeepers and small ICBMs should have superior overall reliability than the SICBMs alone.

Responsiveness to Enemy Countermeasures: Peacekeeper/SICBMs or SICBMs Only?

To counter the mobile Midgetman the Soviets are likely to seek ways to barrage attack the SICBM's entire operating area. If U.S. small mobile ICBMs were dispersed widely enough, this might require the Soviets to fractionate their forces in order to cover such broad area targets with sufficient overpressures to destroy them.

If the SICBMs operating area is sufficiently restricted, the U.S.S.R. might put at risk the entire

mobile missile force. The 12,000 sq. miles on government reservations tenatively allocated to the small
missile may not be sufficient to insure survivability.
Helpful U.S. responses to the Soviet barrage threat
might include increasing SICBM's operating area, maintaining undisclosed operating patterns, and/or retain-
ing the option of leaving the reservations during a
crisis.

The tandem deployment of the Peacekeeper in silos
and the small mobile missile could complicate the
Soviet ability to devise countermeasures. The Peacekeeper in silos, particularly if those silos are superhardened, creates incentives for the Soviets to deploy
large, high-yield warheads. Midgetman, on the other
hand, creates incentives for a large number of smaller
warheads. This two-way pull between fractionation and
defractionation would pose a severe challenge for the
likes of Defense Minister Ustinov, who must deal with
the real world concerns of finite resources.

U.S. ICBM modernization also compels the Soviets
to look to the survivability of their own ICBM forces.
Soviet defensive countermeasures could also include an
increased emphasis on mobile missiles to augment their
silo-based ICBMs. The mobile SS-20 IRBM has been
deployed widely on the European and Asian borders of
the Soviet Union. Also, flight tests are well underway
for a new mobile ICBM, the SS-X-25. As the Scowcroft
Commission argued, U.S. and Soviet moves to mobile
ICBMs should be stabilizing and shifting a large
portion of Soviet ICBMs to more stable, less vulnerable
mobile missiles would simplify the process of reducing
the overall number of warheads, thereby contributing to
arms control.[16]

Another potential countermeasure to the
Peacekeeper/SICBM tandem could be a Soviet breakout
from the ABM treaty and increased emphasis on ballistic
missile defense. Even an imperfect BMD could have a
significant impact on the overall balance. If selected
critical C^3I bunkers and missile silos were protected, U.S. retaliatory ICBMs would be forced to maneuver
to penetrate the Soviet defenses. Soviet BMD could
require the Peacekeeper and small missile to be MARVed
if they are to reach and destroy their targets. Front-
end changes to Peacekeeper could probably be accomplished if it were down-loaded. Because of weight
constraints, MARVing the small missile might prove
impossible or impractical. The two-missile tandem
clearly provides greater potential U.S. responsiveness
to Soviet countermeasures.

Crisis Stabilizer: Peacekeeper/SICBMs or SICBMs Only?
Fixed-silo ICBMs on both sides have become more
vulnerable to rival ballistic missile forces as those
forces become increasingly accurate. This condition

could be changed in the future by superhardening of silos, ballistic missile defenses at ICBM sites, or by deep reductions in counterforce weapons at START. However, even if Soviet ICBMs remained in their present silos and became increasingly vulnerable to U.S. ballistic missiles (Peacekeeper, Minuteman III, the small ICBMs, or Trident D-5), this would still not constitute a credible first strike capability on the part of the United States.

Just as Soviet warplaners must consider the U.S. triad when considering a first strike, so too will U.S. warplanners be forced to face the Soviet triad of strategic forces. The ability to launch strikes against Soviet ICBMs would be of limited value if submarines and bombers were not targeted simultaneously. But, as is the case with the U.S. triad, one element cannot be targeted without giving warning and time to retaliate to the remaining two elements.

Neither the tandem deployment of MX/SICBM or of small ICBMs alone would constitute a credible first strike force against high-value Soviet hardened targets (ICBM silos, launch control centers, or leadership shelters) much less a viable threat against Soviet strategic bombers or ballistic missile submarines at sea.

Were both Peacekeeper (100 ICBMs, 1000 warheads) and 1000 small ICBMs (1000 warheads) deployed, this would still be insufficient to cover half the very important hard targets that are located at fixed sites in the Soviet Union even with a one-on-one allocation of warheads to targets. (The Soviet Union has 1400 ICBM silos, over 300 launch control sites, and thousands of other storage and leadership sites.)

The deployment of 1000 single-warhead small ICBMs, if deployed without Peacekeeper ICBMs, would obviously pose no independent first strike threat either to Soviet ICBMs or to the majority of the other Soviet hardened assets. Normally, a two-on-one ratio of warheads to each hardened target would be necessary to get a high probability of destruction.

The tandem of MX/SICBMs would create the greater deterrent threat to the highest values of the Soviet leadership without, at the same time, constituting a comprehensive first strike capability.

National Will: Peacekeeper/SICBM or SICBMs Only?

There is a legitimate question over whether the "small ICBM only" alternative would adequately preserve the perception of U.S. national will and determination. Unilaterally cancelling the Peacekeeper as it begins the production phase stands the chance of sending the wrong message to our enemies as well as our allies, particularly in the face of a Soviet strategic buildup which has continued without pause. Former

Secretary of Defense Harold Brown concluded that the Peacekeeper is a unique symbol of American national will, the importance of which far exceeds its military value. There is, he concluded, no alternative to the Peacekeeper:

> Recall that we said in the early 1970s that we would modernize with a new missile in the late 1970s. In the mid-1970s we said that we would do so in the early 1980s, and in the late 1970s that we would in the mid-1980s. We have failed so far to do any of those things, even while the Soviets were deploying over 600 new ICBMs, each with a payload equal to or greater than that of MX, and with accuracies now matching those of the most accurate U.S. ICBMs.
>
> To say that the United States will modernize in the early 1990s with a small single-warhead missile will just not be believable. The Soviets would be justified in calculating that any new U.S. ICBM system will be aborted by some combination of environmental, doctrinal, fiscal, and political problems. Our allies in Europe might balk at deploying new U.S. intermediate range missiles on their territory.[17]

Timeliness of Deployment: Peacekeeper/SICBM or SICBMs Only?

The timeliness of the Peacekeeper has never been challenged. Because it is already in production it is the timely response to the Soviet buildup. But, critics of the Peacekeeper argue that if the SICBM can be produced rapidly, the requirement for the Peacekeeper fades. They charge that the Scowcroft Commission's timetable for the SICBM artificially extends a relatively straightforward production process. By establishing "nonexistent" barriers, they contend that the Administration has prolonged by several years the simple task of combining existing technologies. This, they believe, insures that the SICBM will not be ready before the early 1990s and therefore reinforces the need for the Peacekeeper as an interim system.[18]

While the Scowcroft Commission argued against a crash program to speed development, critics contend that this is precisely what is needed. They contend that in the past crash programs have met with considerable success and "storming" the small ICBM problem might yield the same results. Senator Glenn put it this way when he said:

> I simply refuse to believe that we cannot have a mobile force in service in less than the 10 years

projected by the Commission. In the early 1960s
our nation developed the Polaris submarines and
missiles, a truly revolutionary concept, in 3
years from beginning to deployment of the first
submarine. The small mobile ICBM is by comparison
a simple task. We can and should have the first
units deployed in 4 to 5 years, if we are willing
to cut the red tape and bureaucratic obstacles.[19]

The Scowcroft Commission responded to this argument by pointing out that the deployment of both the
Peacekeeper and small ICBM is necessary because a
missile with Peacekeeper-like capabilities is required
now, and because the production of the small ICBM cannot be accelerated without undesirable compromises in
operational effectiveness.

The Scowcroft Commission recognized that the
specifics of the small ICBM are subject to great variations depending on changes in Soviet strategic programs
and general behavior, Congressional funding, the course
of arms control negotiations, and the progress of our
own research and development. Nevertheless, the Commission set 1992 as an IOC target date.

The Small Missile Independent Advisory Group, the
Schreiver Group, established a timetable which concur-
red with the Scowcroft estimate.[20] The Schreiver
Group's schedule begins with a rough concept that has
not yet been studied and leads it to deployment of the
entire system in nine years.

In terms of other weapons programs, the pace of
the SICBM program is ambitious. Recall, for example,
that the program leading to the Trident D-5 SLBM was
begun in 1975 but the missile will not be deployed
until 1989.[21]

Despite the work that has already been done, the
Small ICBM is still a concept, not a missile. Regardless of its ultimate design specifics, the missile will
borrow heavily from technology developed in the Peacekeeper program. Guidance, propulsion, C^3I, and warheads will all be derivatives of Peacekeeper R&D. Much
of this research is new and as yet unfinished.

Accelerating the SICBM's timetable beyond the
recommendations of the Scowcroft Commission and the
Schreiver Group could force design choices which might
compromise the quality of the missile. Some of the
technology envisioned for the missile is already in
existence, but other key systems have yet to be designed. For example, the hardened mobile launcher is only
in the most preliminary stages of design.[22] There
is at present no way of judging with certainty how
effective a mobile launcher will be. A series of tests
and competitions has been scheduled to validate the HML
concept and to determine the most reliable launcher.

The guidance system is another component which would suffer if the development process were cut short. Unfortunately, the AIRS guidance system on the Peacekeeper ICBM cannot be incorporated into the SICBM without modifications. This "beryllium ball" guidance system weighs 450 lbs., 150 lbs. more than the small missile will be designed to carry. Efforts are underway to develop a lightweight model, but AIRS precision is such that a microscopic particle of dust is sufficient to upset the system's accuracy. Clearly, development of a very accurate SICBM is not a simple or fast process.[23]

For a crash SICBM program to be successful, it would require a single-minded U.S. purposefulness which precludes much further debate. It is unlikely that Congress and the Executive branch could keep the SICBM above further debate for the several years necessary to build and deploy the missile. More than likely, a crash program would be subject to the fluctuations in funding and fiscal belt-tightening which accompany major programs. Any change of administrations would produce the inevitable policy review. A crash program marked by speed-ups, slow-downs, and change in funding would ineffectively use limited resources, and may well delay the SICBM's deployment.

Also, a crash program requires the R&D phase and the production phase to proceed simultaneously. Problems that were discovered during R&D would have to be corrected in the units already produced. As John Medalia of the Congressional Research Service has observed, "Such a program would accept higher technical risk. Indeed, attempts to accelerate the IOC could backfire and ultimately delay the program."[24]

Tandem deployment eliminates the need for an accelerated IOC for the small missile. The Peacekeeper is already in production and is scheduled to enter service in 1986. This deployment will immediately address the asymmetry of forces which is the immediate reason for modernization. Without the pressure of a crash program, the SICBM will be able to make the fullest use of existing technology. As a result, the SICBM can be made more responsive to the long-term requirements of deterrence.

Penetration of Enemy Defenses: Peacekeeper/SICBMs or SICBMs Only?

The Scowcroft Commission warned that:

As Soviet ABM modernization and modern surface-to-air missile development and deployment proceed -- even within the limitations of the ABM treaty -- it is important to be able to match any possible Soviet breakout from that treaty with strategic forces that have the throwweight to carry sufficient numbers of decoys and other penetration aids.[25]

The SICBM will operate under severe weight restrictions. To maintain its 30,000 lb. target weight both the guidance system and warhead have already been trimmed. It is uncertain that the SICBM will have the throwweight to carry penetration aids. The Peacekeeper, on the other hand, has the throwweight to carry whatever such penetration aids as may be necessary. Thus, dual deployment offers an advantage over the "SICBM only" in that it allows for the inclusion of Peacekeeper penetration aids.

Arms Race Stability and Cost Control: Peacekeeper/SICBMs or SICBMs Only?

It is possible that deployment of 100 Peacekeeper ICBMs will provoke no appreciable Soviet arms building responses beyond their current and planned programs. After all, 1000 MX warheads could not target all 1400 ICBM silos, as well as the several hundred additional hardened Soviet leadership bunkers, Soviet launch control centers, and Soviet nuclear weapons storage sites and several thousand other important Soviet targets.

Moreover, it is not at all clear that the Soviet Union could do much more in strategic weapons development if it tried under present peacetime conditions. Already, vast resources are being poured into new Soviet bombers, missiles, submarines and warheads. The flood of research, development, and deployment on new weapons is sustained by a Soviet military bureaucracy that has budgetary clout and political support. The current pace is already an enormous strain on the Soviet economy. New U.S. ICBMs, (Peacekeeper alone, SICBM alone, or both together) may create a mild stimulus akin to a jockey's whip on a horse already in full stride, but the difference in velocity of Soviet programs probably will not be all that different from what would have taken place anyway.

One of the basic aims of U.S. arms control policy is to reduce the costs of the arms race to U.S. citizens. Life cycle and operations and maintenance costs (per warhead) of the small ICBM program would significantly out-distance the expenses of silo-based Peacekeeper ICBMs.[26] Therefore, if the U.S. planned to deploy a given number of ICBM weapons, a tandem force of Peacekeepers and small missiles would be cheaper than an equivalent number of warheads mounted just on single-warhead small ICBMs.

Arms Control Leverage: Peacekeeper/SICBMs or SICBMs Only?

Without the pressure of Peacekeeper ICBM deployments, the Soviet Union will have little incentive to rejoin the START negotiations and bargain reductions in force. Peacekeeper ICBMs carry a potent threat to Soviet power and authority and would do it years sooner than would a small ICBM program alone.

The Soviet leaders are not philanthropists. They will not agree to greater cuts in their ICBM forces than the U.S. is willing to make. The U.S. needs to match Soviet ICBM countersilo strength (by deploying Peacekeeper, then SICBMs) before it could ever realistically expect Soviet reductions in SS-18s and SS-19 as a result of resumed START negotiations.

Electing to abandon Peacekeeper and to add only small mobile ICBMs at some later date to the present force can only diminish U.S. bargaining leverage in future arms control negotiations with the Politburo. The U.S. threat to their force would diminish and they might be tempted to stall any further negotiating progress to see if U.S. domestic politics would cut other U.S. strategic deployments without requiring a Soviet concession at the START neogitations.

Conformity With Present Agreements: Peacekeeper/SICBMs or SICBMs Only?

The deployment of small mobile ICBMs alone would not violate the terms of the SALT II Treaty if either were to occur after December 31, 1985 -- the date when SALT II is to expire.

In the unlikely case of SALT II still being observed after December 31, 1985, the Treaty would prohibit the testing of a second "new type" of ICBM such as the small mobile. However, it is conceivable that the United States will not continue to adhere to SALT II due to violations of the pact by the Soviet Union.

SALT II also prohibits launchers that could be rapidly reloaded, any change of silo locations, additional silos under construction, and any increase in silo volume that exceeded 32 percent. So long as both Peacekeeper and the SICBM programs stayed within those limits there would be no violation of the SALT II treaty limits.

The introduction of a large number (e.g., 1000) small ICBMs would cause the United States to exceed the SALT II limits on overall strategic nuclear delivery vehicles (SNDVs) by a wide margin. This would be true also of any future START agreement that used the SNDV or launcher as the basic measure of strategic merit. Therefore the acceptance of the last Soviet START proposal, which is basically a modification of the SALT II

Treaty, would rule out deployments of large numbers of SICBMs, or would force a major drawdown in the numbers of permitted U.S. warheads if the U.S. decided to proceed with SICBMs because it would need to replace MIRVed missiles with single-RV systems to stay within the START launcher limits. A change of the measure of merit in future arms control agreements from counting launchers to counting warheads deployed would solve the problem for the U.S. and still allow both sides to agree on arms limits.

Unless the counting rules of any future START agreement were changed from past SALT agreements, both the SICBM-alone option and the Peacekeeper/SICBM tandem option would quickly run afoul of possible future START ceilings. Only the Peacekeeper-alone deployment option could pass through the START hoop if launchers were what were counted and restricted. However, the U.S. Government is likely to continue to insist upon arms control counting rules that would permit both Peacekeeper and small ICBM deployments in adequate numbers or no new agreement would be signed.

Verifiability: Peacekeeper/SICBMs or SICBMs Only?

Peacekeeper ICBM deployments in fixed silos can be easily monitored using national technical means (NTMs) of verification. Cheating on, or compliance with, ongoing arms control ceilings on fixed launcher numbers can be readily ascertained in a timely manner.

On the other hand, deployment of small mobile ICBMs becomes a greater verification problem the more dispersed or disguised the force. SICBMs deployed in certain specific, if large, operating areas would lend themselves far more readily to inspection and counting than if these same small ICBMs were deployed on U.S. highways in trucks with no observable differences from other trucks.

The disguised off-based deployment of SICBMs would present a severe challenge to verification and to negotiating verifiable arms control agreements.

However, it should be noted that the Soviet Union has a far easier time in gathering intelligence on U.S. force levels due to the openness of our society. This should allow them to confirm by other intelligence means what may not be ascertainable by national technical means of verification. Moreover, the main U.S. concern about verification is whether or not we are setting a precedent that the Soviets will follow (mobile ICBMs deployed off-base) and which might make us blind to their deployment levels. Throughout the past four decades of U.S.-Soviet negotiations it has continually been the United States that has shown the most concern about verification questions. The U.S.S.R. has been relatively unconcerned. Therefore the real U.S. concern in choosing an ICBM deployment

mode is to select one that makes sense from a military
and stability standpoint and which permits adequate
verification if the Soviets should mirror our ICBM
deployment modes.

Conclusions

In their efforts to see the Peacekeeper cancelled,
critics have held up the small mobile ICBM as an ideal
ICBM capable of handling all future deterrence
requirements. The small ICBM, we are told, is less
vulnerable and more stabilizing than Peacekeeper,
conducive to arms control and can be made available
just as quickly.

However, a point by point examination of all of
the important characteristics of an ideal ICBM force
suggests that the small mobile missile by itself, is
not the panacea that it is sometimes portrayed to be.
Mobility may contribute to the survivability of small
ICBMs, but it also limits its realiability, retargetability, enduring C^3I, its ability to penetrate
active defenses, and the ability to verify whether an
ICBM force is in compliance with arms control
agreement. In short, the small ICBM has liabilities as
well as assets.

Dual deployment of the Peacekeeper and the small
ICBM accomplishes what neither system could do individ-
ually. Together the two missiles address the necessary
requirements of an ideal ICBM force. A synergistic
relationship is established wherein the virtues of each
weapon contribute to overall capability. The net
result is a responsive, highly survivable ICBM force
which could be fielded quickly and that would hold at
risk the high value hardened targets of the Soviet
state. Such a tandem force composed both of Peacekeeper and small ICBMs should be very healthy for near-
term and long-term U.S. national security. Two new
ICBMs appear more helpful than either type alone. For
a comparison see Table 6.2.

Table 6.2
RATING OF ICBM OPTIONS

Desired Traits	Peacekeeper Alone	Small ICBM Alone	Tandem Deployment	Ideal ICBM Force
Reduces Risk of War	Good	Good	Excellent	Excellent
Survivability	Poor/Good*	Fair/ Excellent	Good/ Excellent	Excellent
Accuracy	Excellent	Excellent	Excellent	Excellent
Reliability	Excellent	Fair/Good	Good/ Excellent	Excellent
Enduring C^3I	Fair	Poor/Fair	Fair	Excellent
Prompt & Flexible Response	Excellent	Good	Excellent	Excellent
Responsiveness to Countermeasures	Fair	Good	Excellent	Excellent
Crisis Stabilizer	Fair	Good	Good	Excellent
Symbol of National Will	Excellent	Poor/Fair	Excellent	Excellent
Timeliness of Deployment	Excellent	Fair	Excellent	Excellent
Penetrates Enemy Defenses	Excellent	Fair	Good/ Excellent	Excellent
Arms Race Stability and Cost Control	Good	Good	Poor/Fair	Excellent
Arms Negotiation Leverage	Excellent	Fair	Execellent	Excellent
Conformity with Present Agreements	Excellent	Good	Fair	Excellent
Verifiability	Excellent	Fair	Fair/Good	Excellent

* Depends on U.S. strategy (e.g., launch-under-attack) or on how the uncertainities play out.

NOTES

1. President's Commission on Strategic Forces, Report, April 6, 1983, p. 8.

2. Ibid., p. 14.

3. Tom Wicker, "Flight 7 and the MX," The New York Times, September 9, 1983.

4. Daniel Patrick Moynihan, "Reagan MX Plan Commits U.S. to First Strike Policy," Long Island Newsday, July 26, 1983, p. 39.

5. Report, op. cit., p. 13.

6. Ibid., p. 14.

7. Significant Congressional Hearings dealing with the Scowcroft Commission were held by the Senate Foreign Relations and Armed Services Committees, and by the House Armed Services and Appropriations Committees. Major debates also took place on the floor of the House and Senate in May, July and November of 1983.

8. See, National Institute for Public Policy Mono-graph, "U.S. National Security Policy Implica-tions of Superhard Silos," March 1984.

9. Testifying before the Senate Armed Services Committee, March 6, 1984, General Fornell explain-ed that a cap of 100 psi had been set on the HML, but that industrial sources were convinced that 200 psi was feasible.

10. See, for example, "Pentagon is Studying Mobile Garage for Minuteman to Survive Surprise Attack," New York Times, January 7, 1970, p. 10.

11. Testimony of Dr. Thomas Cooper before the Senate Armed Services Committee, March 6, 1984.

12. Report, p. 10.

13. See, for example, Barry R. Schneider, "Invitation to a Nuclear Beheading," Across the Board, July-August 1983, pp. 9-16.

14. Small Missile Independent Advistory Group, Report, September,1983, pp. 10-11.

15. Modernizing U.S. Strategic Offensive Forces, The Administration's Program and Alternatives, (Washington, D.C.: Congressional Budget Office, May 1983), p. 84.

16. Report, op. cit., pp. 22-23.

17. Harold Brown, "Statement on the Report of the Presidential Commission on Strategic Forces," April 11, 1983, p. 6.

18. "Time, Cost and Base for Small Missiles Under Vital Debate," New York Times, September 7, 1983, p. B9; see also, "Midgetman Questions Shadow MX," Washington Post, June 4, 1983, p. 11.

19. Congressional Record, May 24, 1983, p. S7324.

20. Small Missile Report, op. cit., pp. 15-16.

21. Thomas Cochran, William Arkin, and Milton Hoenig, Nuclear Weapons Databook, (Cambridge, Mass.: Ballinger, 1984) pp. 145-156.

22. HML technology is one area where small increments in performance may yield dramatic results. For example, a one megaton warhead could destroy the SICBM anywhere within an area of 10 square miles if the HML were hardened to 20 psi. If it were hardened to 40 psi, the same warhead would be lethal only over 3 square miles. John Medalia, "Small Single-Warhead ICBMs: Hardwar, Issues and Policy Choices," Congressional Research Service Report, 83-106, May 26, 1983, p. 27.

23. James W. Canan, "A Full-Bodied Triad," Air Force, February 1984, pp. 51-52.

24. John Medalia, op. cit., pp. 20-21.

25. Report, op. cit., p. 17.

26. The Reagan Administration has not attached a cost estimate to the Small ICBM program, explaining that the missile is in the early stages of concept definition, and that costs will be dependent on a number of variables which have yet to be decided. However, the Congressional Budget Office has estimated the 20 year life-cycle costs of 1000

small ICBMs to be $107 billion. See, <u>Modernizing</u>
<u>U.S. Strategic Offensive Forces: The Administra-</u>
<u>tion's Program and Alternatives,</u> p. 58.

7
Soviet Uncertainties in Targeting Peacekeeper

Barry R. Schneider

Introduction

For the past decade U.S. defense analysts have charted the progress the Soviet military has made in strategic rocketry. The capabilities of Soviet intercontinental ballistic missiles have improved by an order of magnitude. The Soviet SS-18 and SS-19 ICBMs have the number of warheads, and appear to have the accuracy, yield, and reliability to endanger the United States silo-based ICBM force in a first strike attack. As a result, as this Soviet force has grown in numbers and accuracy the U.S. ICBM force has been said to be entering a "window of vulnerability" in the 1980s.

Solutions to this vulnerability problem have centered around: (1) new ICBM basing modes and (2) new and more capable U.S. ICBMs, the Peacekeeper and the new single-RV Midgetman ICBMs.

Present DoD plans call for one hundred Peacekeeper missiles to be built and based in existing Minuteman silos, replacing a portion of the less capable Minuteman IIIs at Warren AFB in Wyoming and Nebraska. Later this force may be augmented by mobile single-warhead Midgetman ICBMs deployed widely enough to alleviate the ICBM vulnerability problem.

Some have questioned the wisdom of basing Peacekeeper missiles in present Minuteman silos because this does not appear to address the "window of vulnerability" problem posed by the SS-18s and SS-19s.

There are several basic responses to those who question U.S. ICBM silo-basing:

-- ICBM vulnerability is partially addressed by the triad synergism. (See Chapter 2.) Even vulnerable silo-based ICBMs would serve to protect the bomber leg of the triad by creating attack coordination and timing problems for the Soviet warplanners.

-- Silo-basing of new ICBMs may be the only available deployment mode that is acceptable simultaneously from a military, political and economical standpoint. Furthermore, the new Peacekeeper missile provides more deterrent effect and retaliatory punch than present silo-based ICBMs.

-- Present silos might be significantly upgraded in hardness in the future and/or actively defended, thereby improving their survivability.

-- U.S. ICBMs in silos might be considerably safer than is popularly conceived because of major uncertainties about the effectiveness of a Soviet first strike attack.

The President's Commission on Strategic Forces recommended Peacekeeper missiles in silos, because it was an adequate and acceptable way of basing the ICBMs at an early date.
The Scowcroft Commission concluded that:

...the vulnerability of such silos in the near term, viewed in isolation, is not a sufficiently dominant part of the overall problem of ICBM modernization to warrant other immediate steps being taken....This is because the mutual survivability shared by the ICBM force and the bomber force in view of the different types of attacks that would need to be launched at each. To deter (Soviet)...surprise attacks we can reasonably rely both on our other strategic forces and on the range of operational uncertainties that the Soviets would have to consider in planning such aggression...[1]

After analyzing the uncertainties that face Soviet warplanners, should they be planning a first strike attack against the United States, a set of basic conclusions emerge:

(1) U.S. ICBMs may be far less vulnerable to destruction by a surprise attack carried out by Soviet ICBMs than is commonly supposed.

(2) The uncertainties facing Soviet warplanners about whether or not a surprise attack against ICBMs will succeed are massive.

(3) A Soviet Union, uncertain about whether or not its forces can knock out United States ICBMs, ought to be deterred from attack given

the massive penalties for even a slight failure.

Before discussing whether or not the Soviets could acquire a high confidence capability to launch a nuclear knockout of the U.S. silo-based ICBM force, it is important to put this discussion in context. The first and most obvious point is that Peacekeeper ICBM vulnerability is not the same as triad vulnerability. That is, even if the Soviet forces should acquire a theoretical capability to destroy every American ICBM in its silo, the United States would still retain enough retaliatory capability with bombers and SLBMs to to destroy a significant portion of Soviet military bases, military capabilities, defense industries, transportation and communications systems, economic base, and Soviet population.

Sit for a few moments in the position of a Soviet warplanner assigned to work out the results beforehand of a possible future Soviet attack against the U.S. silo-based ICBM force. The first fact that would emerge from such an analysis would be the unpredictability of the outcome. Second, this Soviet warplanner should see that the probability of a devastating response by the defender's ICBMs is far higher than the chance of a Soviet attack that could destroy 90 percent or more of the U.S. ICBM force.

Clearly large uncertainties would exist today in predicting the number of likely ICBM survivors in the United States after a Soviet SS-18 and SS-19 nuclear first strike. The size of the uncertainty would be a function of uncertainties in:

(1) The United States strategy adopted when warned of the impending attack or implemented when under attack.

(2) The scenario leading up to the attack, the mix and number of forces available on each side, and the alert status of forces at the time of the attack.

(3) The performance of Soviet forces and weapon systems in operational conditions.

(4) The effectiveness of Soviet attack strategy and their coordination of attack forces.

(5) The United States leadership, C^3I, and strategic offensive force vulnerability or survivability.

(6) Local factors influencing the battle outcome (nuclear effects, geology, silo hardness, missile

hardness, atmospheric conditions, and weather effects).

Note Table 7.1 for a summary of the kind of uncertainties facing a Soviet warplanner contemplating a first strike attack against U.S. ICBM forces.

Table 7.1

The Soviet Warplanner's View of Major Uncertainties
In Targeting Peacekeeper ICBMs

U.S. Strategy Uncertainties:

-- U.S. launch-on-warning?
-- U.S. launch-under-attack?
-- U.S. rideout strategy?
-- Presidential predelegation to U.S. forces?
-- U.S. countermilitary response?
-- U.S. countervalue response?

Soviet Weapon System Uncertainties:

-- Operational versus test accuracy?
-- Warhead yield?
-- Weapon system reliability?
-- Height of burst/fuzing?
-- Launch rate?
-- Attack coordination and timing?
-- Availability of forces?
-- Command and control effectiveness?
-- Support system effectiveness?

Target-End Uncertainties in CONUS:

-- ICBM silo hardness to blast? Radiation?
-- ICBM hardness to blast? Radiation?
-- Precise knowledge of target location?
-- National Command Authority vulnerability?
-- Strategic C^3I vulnerability?
-- Local geology and blast effects on silos?
-- Target-end weather effects?
-- Possible fratricide effects?
-- Impact of superhardening on outcomes?
-- Impact of predelegation to commanders on outcomes?
-- Clandestine U.S. forces?

Soviet Uncertainties About U.S. Strategy

Clearly the greatest uncertainty facing any Soviet warplanner would be whether or not the United States would:

(1) ride out the attack before attempting to respond; or

(2) launch under attack as soon as the first detonations began and were confirmed; or

(3) launch on tactical warning of the attack; or even

(4) preempt the Soviet attack upon receiving high confidence strategic warning that such an attack was about to take place.

The postulated vulnerability of U.S. silo-based ICBMs to super-accurate Soviet SS-18s and SS-19s argues against riding out a Soviet countersilo attack. The vulnerability of silos is compounded by the vulnerability of U.S. National Command Authorities and of critical command, control, communications, and intelligence assets to Soviet attacks.

The Soviet warplanner must ask himself what the U.S. might do to offset vulnerabilities to ICBMs, the NCA, and C^3I assets. One logical answer might be to launch the force on tactical warning of the attack. The Soviet warplanners could not know whether or not such an order might even, in certain contingencies, be predelegated by the President to the SAC commander or to the U.S. Air Force General who flies aboard "Looking Glass" at the moment of confirmation of the Soviet attack in progress.

A United States "launch-on-warning" (L-O-W) capability is clearly within our competence to perform. Much more difficult would be a "launch-under-attack" (L-U-A) capability. With the present U.S. ICBM force at full alert, over 2000 Minuteman and Titan warheads could be launched on warning and be on their way to targets in the Soviet Union before the first Soviet warheads could be detonated on American ICBM silos. The addition of 100 Peacekeeper ICBMs, as planned, would mean that nearly 3000 U.S. ICBM warheads could be fired before the Soviet attack arrived. The uncertainty of whether or not the U.S. would launch-on-warning is the U.S. trump card that could thwart the effectiveness of any Soviet attempt to destroy U.S. ICBMs.

As former Secretary of Defense Harold Brown has pointed out:

The Soviets would have to consider the possibility
that our Minuteman (and future MX) missiles would
no longer be in the silos when their ICBMs
arrived. We have not adopted a doctrine of launch
under attack, but they surely would have to take
such a possibility into consideration.[2]

The Soviet warplanner would have major uncertainties about the amount of strategic and tactical warning
received by the United States about the impending
attack.

When the Soviet planner tries to calculate some
future outcome of a Soviet-U.S. nuclear exchange, they
must make significant (and perhaps wrong) assumptions
about how the opening crisis would unfold, and what
U.S. intelligence could glean from Soviet preparations.
Then he would have to predict the U.S. reaction in the
preattack, transattack, and postattack periods. He
would have to make major assumptions about the ability
of the U.S. leadership to survive and its capability to
order retaliatory attacks in all phases of the attack.
The Soviet planner would have to decide whether or not
the U.S. military has the predelegated authority to
order nuclear counterattacks if the NCA is disabled by
the attack.

U.S. ICBMs are continually kept at almost a 100
percent alert rate even in normal noncrisis periods.
Peacekeeper, when fully deployed will provide an additional 1000 highly accurate reentry vehicles that can
be maintained day-to-day on maximum alert just moments
away from launch if necessary.

Soviet Uncertainties About Soviet Weapon Effectiveness

The Soviet warplanner also faces major uncertainties regarding the performance and effectiveness of
Soviet weapons. Some of these include uncertainties
in:

-- accuracy   -- height of burst
-- systemic bias error   -- timing of detonation
-- yield of warhead   -- launch rate and timing
-- reliability   -- command and control
-- availability   -- support system
-- range   effectiveness

Accuracy Uncertainties

The greatest problem in forecasting the results of a first strike attack is the uncertainty in forecasting missile accuracies during an actual attack. Accuracy is the single most important variable in determining the success or failure of a countersilo attack. Both U.S. and Soviet officials face the same kinds of technical uncertainties about their missile performance. While we are not privy to the internal Soviet debate over missile accuracy, the internal U.S. discussion about missile accuracy or inaccuracy is instructive.

There are two schools of thought expressed. The official school of thought is that ballistic missiles are extremely accurate, and inaccuracies can be predicted and accounted for. The second is the skeptical school of thought that suggests that test experience with ICBMs may be very different from a operational experience. The skeptics question the accuracy and countersilo potential of ICBMs if used in wartime conditions. To understand this debate it is instructive to look at how we attempt to gauge ICBM accuracy on the test range.

Numerous possible sources of inaccuracy can make a missile reentry vehicle miss and inflict less-than-expected damage against a target. Consider, for example, the problem of destroying a target 5,500 miles away with your first shot ever at that target. An adjustment process also goes into test firing and correcting the accuracy of missiles. The Soviet Union, for example, test-fires ICBMs from Tyuratam in the Western U.S.S.R., or along a Northeast flight path to an impact area on the Siberian Kamchatka Peninsula or into a target zone in the Northern Pacific Ocean. The United States tests its ICBMs by firing them from Vandenberg AFB in California, westward across the Pacific to a target area in the lagoon at Kwajalein Atoll 3800 miles away. The first missile shots tend to go widest of the mark until adjustments are made to correct for the systemic bias of the test launcher and the missile. This "bias" factor is the difference between the calculated center of missile impact points on the test range and the actual target aimpoint.[3]

The aim of repeated missile test flights is to bring the shot pattern more nearly around the intended target (i.e., to correct for the bias) so that later missile shots hit close enough to the intended target to have destroyed it had the warhead been detonated.

However, test-range results might be more accurate than wartime operational shots against real targets. We have discovered in tests that the bias of a missile can change with the azimuth of the launch in sometimes unpredictable ways. That is to say that Soviet ICBMs launched over the North Pole at targets 5,500 miles away in the United States might be less accurate the first time the attempt is made because the conditions along each separate flight path vary.

Of course, during the research, development, engineering, and test phases of a missile program, scientists in the Soviet Union and the United States eliminate many of the potential accuracy errors in the system. For example, future Peacekeeper missiles will be extensively tested and constructed so that daily variations in atmospheric density, anomolies in the earth's gravitational field, and changes in winds, temperature, and precipitation can be offset to maintain better accuracy. Inertial guidance systems will be tested against vibrations, shock of reentry, and developed to make minute flight-path corrections during the boost phase of powered flight.

Some officials believe that RDT&E programs can remove all the bugs from a missile and that excellent accuracy, equivalent to test range results, will be the operational result if the ICBMs are ever fired in anger. As Seymour Zeiberg, former Deputy Under Secretary of Defense for Research and Engineering for Strategic and Space programs, has said:

> It may take around twenty test flights before you get rid of all the problems, but finally you do. Along the way, you find all the problems. The idea of some new distortion showing up just doesn't happen if you build and test the guidance programs properly. There are no unknowns. There are no surprises.[4]

This, of course never would be provable until a nuclear strike or counter-strike takes place. Physicist Richard Garwin is not so sure that there would be no surprises. He concedes that "Every factor you can think of has been accommodated"[5] and adds that "It is the things you don't think of that cause the trouble."[6]

Dr. Garwin notes that:

> ...every time you fire a new-model missile over the same range or the same missile over a slightly different range, the bias changes. Sometimes it is greater, sometimes it is smaller, but it never has been calculated beforehand.
>
> So you have to go back to readjusting the gyros and so on, to try and eliminate the novel bias. But if we were firing operationally, both we and the Russians would be firing over a new

range in an untried direction—north. And a whole
new set of random factors would come into play
—anomalies in the earth's gravitational field,
varying densities of the upper atmosphere or
unknown wind velocities. They may adjust and
readjust in testing and eventually they might feel
sure that they have eliminated the bias. But they
can never be absolutely certain. We certainly
cannot be; and although we are less well informed
about the Russian ICBM test program than our own,
there is no reason to suspect that they are any
more successful than we are at dealing with the
problem. If you cannot be sure that you would be
able to hit the enemy's silos, then there is no
point in even trying—because the idea is that one
side could wipe out the other's missiles before
they are launched in a first strike.[7]

U.S. and Soviet experts know that inaccuracy in a
missile system can come from:

1) imperfect prediction of the effects of
   gravity on the missile along a new previously
   untested flight path.

2) errors caused by malfunctioning or imprecise
   inertial sensing and guidance systems (e.g.,
   gyroscope or accelerometer drift and mis-
   alignment, a bending of the stabilizing
   platform, rocket motor magnetic field
   effects, thermal gradients, and voltage
   fluctuations).

3) malfunctions or errors in calculations performed on board the missile during flight.

4) malfunctions and errors during separation of
   the first stage, second stage, and MIRV bus
   from the reentry vehicles.

5) misestimates of conditions (wind, air
   density, temperatures, weather, thermal
   layers) at either the launch end or reentry
   end of the missile trajectory.

6) Secondary sources of error (probably very
   small) such as those caused by:

   - sun/moon gravitational pull.

   - precession and nutation of the earth.

- earth wobble.

- variation in accuracy of measuring earth's surface and topography.

- thrust cutoff timing problems.

- solar wind effects.

- variations in the earth's magnetic field.

- mapping errors of the target location.

Soviet scientists and officers are likely to be well informed on these sources of ballistic missile inaccuracy. It is not clear whether the same kind of debate over missile accuracy exists in the Soviet Union as in the United States. Nor is it clear whether or not Soviet political leaders are knowledgeable about the technical uncertainties involved in predicting whether a Soviet ICBM reentry vehicle can be guided close enough to explode within a lethal distance from U.S. ICBM silos. Nevertheless, it is clear that U.S. scientists and military officials are divided on the questions of whether or not the U.S. (and by implication, the U.S.S.R.) can field a high confidence, very accurate ICBM force.

If the leaders of the Soviet Union believe, as top U.S. Air Force officials do, that their ICBMs are as accurate as test range results indicate, then this would probably not be a factor that gives them pause when considering a first strike. However, if the Soviet military or political authorities harbor the same misgivings about ICBM accuracy in wartime conditions as U.S. skeptics then they ought to be daunted and deterred by such accuracy uncertainties since inaccurate ICBMs could not carry out a successful countersilo attack.

Nevertheless, it is possible that the officials of the Soviet Rocket Forces and those on the Defense Council within the Politburo share the same confidence as the U.S. Air Force in ICBM precision and accuracy. The official U.S. Air Force view was expressed this way by one expert who concluded that:

What makes us feel comfortable that we can predict what the maximum bias or error will be of a flight over a path we cannot practice -- and that route certainly can be described as a path from the U.S. over the North Pole -- is that we can replicate all the phenomena that we expect to see over that path.

We can replicate them by flying over several paths that are different from that of the North Pole and different from each other. We can see what happens to gravity changes, what happens to magnetic fields at different times of the year, when there are sunspots, when there are not, when the Van Allen Belt is acting in a certain way. We've been doing it for 20 years and that accumulated base of knowledge gives the engineer the ability to say, "I know what's handling that particular problem, what's controlling it and I know how much I don't know." That's important. How much you don't know is the number of feet of bias you might expect because you just cannot measure it or calculate it more accurately.[8]

U.S. Air Force and Defense officials have stated that it is their conviction that the earth's gravitational field -- measured on land and sea at the North Pole and from above by aircraft and satellites gives the U.S. (and Soviet) military an adequate knowledge of the effects of gravity on ICBMs launched over the pole. Notes one expert:

We can be extremely accurate "to quantities and numbers far above what we need in hard target capability on any individual flight path. We've done this, we have measured gravitation, we have taken the flights, we have done the experiments and confirmed them....We measure these variables, obviously at discrete points, and then spatially connect these points or average them over the distance we're concerned with."[9]

As General Robert Marsh, Commander of the U.S. Air Force Systems Command, argues:

One of the popular theories is that we do not know this gravity model with sufficient precision to preclude large bias errors when flying over untried routes. In fact, the product of the universal gravitational constant and the mass of the earth, the relevant parameter, is known to one part in 20 million...the gravitational field in the launch region is carefully measured and accurately modeled in such a way that gravitational errors incurred during the guided portion of powered flight (approximately three minutes) are small and boundable. Furthermore, by observing the orbits of our satellites, we obtain data concerning the shape of the earth and the gravitational field in the region of missile flight.

These data allow us to model the gravity field, thereby eliminating significant impact errors true to the gravitational field along the warhead's trajectory.[10]

The official case -- that ballistic missiles over a polar route are as accurate as advertised -- is reinforced by the fact that the U.S. Navy has fired SLBMs on northern azimuths and have seen no major anomolies that the error budgets had not accounted for. These tests, granted, were not directly over the North Pole but lend support to the official story that missile accuracies can be maintained in polar trajectories.

Nor, according to General Marsh, does the United States have a bias which exceeds the CEP in any of our strategic ICBMs or SLBMs. This is a conclusion based on multiple tests over a wide variety of ranges and azimuths.

A number of scientists and engineers who participate in the U.S. ICBM modernization program believe that U.S. officials (and presumably Soviet officials) have a good grasp of the predicted variance they can expect at different confidence levels in missile performance across the entire force.

Thoroughly tested is the predicted variance in missile accuracy due to change in the range, azimuth, powered flight profile, hardware and software guidance packages, launch region, gravitational field along the flight path, target location, wind velocity, and atmospheric density variations above the target.

While USAF (and Soviet) officials might not know the exact accuracy of an individual missile, they believe they know within certain limits what its CEP will be. Military officials could be less confident in predicting the outcome of a single ICBM shot or of a limited nuclear strike against a small target set than they would be in predicting outcomes of larger scale ICBM attacks where the larger statistical sample would yield more predictable results.

General Marsh believes that many possible sources of uncertainty mentioned by others are so insignificant that they can largely be ignored. He argues that such factors as the gravitational attraction of the sun and moon, the pull of the earth's magnetic field acting on the electric charge built up on the ballistic reentry vehicle, the effect of the earth's "wobble," and the impact of target-end winds and variable air densities are so minor that corrections need not even be made for them.[11] The calculated variance from such effects is probably less than a few tens of feet.[12]

While it is possible that U.S. and Soviet official calculations have exactly captured the forces operating on an ICBM over a polar trajectory, this can never be known with absolute confidence until after a missile exchange has taken place.

A Soviet warplanner would, of necessity, need to question whether his test range ICBM accuracies could be converted into operational wartime accuracies due to two other factors: (1) lack of a statistically significant number of missile flight tests and (2) lack of sufficient accuracy checks over the full range of the missile as required in wartime.[13]

The high cost of ICBMs limits the number of flight tests the Soviet Union (and the United States) is willing to perform. Both countries, due to budgetary restraints, fall far short of testing the number of missiles that could yield statistically significant predictions of CEP and missile bias.[14] None are tested over full flight paths.

Inaccuracies due to extended operational ranges are also predictable say Air Force engineers. True, they agree a wartime launch of an ICBM would take place over a 5500 nm range whereas the distance between Vandenberg AFB and the lagoon at Kwajalein is only 3800 nm. Nevertheless, they believe that they can compute accurately the margin of additional error that accumulates over a longer range.

Nevertheless, CEPs expressed in precise numbers to the hundredth of a nautical mile or precisely to a given number of feet or meters -- if those numbers are not based on adequate data -- can create what philosopher Alfred N. Whitehead termed "the fallacy of misplaced concreteness."

The number of tests of a given missile using the same guidance software, the same warhead materials, and aerodynamics is small. As one expert has observed that the U.S. Air Force conducts "what might be called a small segment of pristine, identical type shots. A half dozen, or a dozen in one case I'm aware of, would represent the kind of numbers we're talking about. It would more frequently be six than 12."[15] Soviet ICBMs are more fully flight tested than their U.S. counterparts, and are tested from operational silos, but uncertainty problems could still be substantial.

A ten percent degradation in SS-18 accuracy, by itself, however, might not be enough to save U.S. silos from destruction given the high estimated accuracy and yield of the SS-18. However, if Soviet leaders understood that there could be a 10 percent error in estimated accuracy combined with other similar errors in estimating weapons reliability, yield, availability, etc., the accumulated uncertainties ought to dissuade Soviet decision-makers from ordering a first strike even in extreme crises. This assumes, of course, that Soviet leaders are fully briefed by their military advisers on the technical uncertainties involved. One

U.S. Secretary of Defense had some advise for Soviet and U.S. leaders on this question.

On March 4, 1974, Secretary of Defense James Schlesinger gave secret testimony, subsequently declassified, to the Senate Foreign Relations Committee on the subject of not placing too much confidence in missile attacks. Schlesinger testified that:

> I believe that there is some misunderstanding about the degree of reliability and accuracy of missiles...It is is impossible for either side to acquire the degree of accuracy that would give them a high-confidence first strike, because we will not know what the actual accuracy would be like in a real world context. As you know, we have acquired from the western test range a fairly precise accuracy, but in the real world we would have to fly from operational bases to targets in the Soviet Union. The parameters of the flight from the western test range are not really very helpful in determining those accuracies to the Soviet Union. We can never know what degrees of accuracy would be achieved in the real world...
>
> The point I would like to make is that if you have any degradation in operational accuracy, American counterforce capability goes to the dogs very quickly. We know that, and the Soviets should know it, and that is one of the reasons that I can publicly state that neither side can acquire a high-confidence first strike capability. I want the President of the United States to know that for all the future years, and I want the Soviet leadership to know that for all the future years.[16]

Uncertainty in Weapon System Reliability

Another important variable in assessing Soviet countersilo capabilities is missile reliability. The Soviet SS-18 suffered 7 failures in its first 29 research and development flight tests.* This meant the SS-18 had a reliability of 76 percent on the test range.[17] The Soviet SS-19 had 2 observed failures in its first 27 tests a weapons system reliability of 92 percent.[18] The Soviet warplanner might worry about some unknown "X" factor that could cause the failure of an entire class of ICBMs (e.g., SS-17s, SS-18s, or SS-19s) against designated U.S. targets. U.S. World War II experience with torpedoes is a

---

* Soviet missiles are also test flown from operational silos after deployment to improve and prove the operational accuracy and reliability of such weapons.

classic example of a standard weapon system that totally malfunctioned during critical early battles. Any warplanner would be less than prudent if somewhere in his planning he did not prepare for the contingency expressed so eloquently in Murphy's Law: "If anything can go wrong, it will."

United States ICBMs are not fired from operational silos but all system components right up to the moment of ICBM engine ignition are tested on a regular basis. Then, selected missiles are pulled from operational silos and transported to Vandenberg AFB where they are test launched out over the Pacific Ocean. The Minuteman force has been on alert for over twenty years and has had reliability/alert rates over 90 percent throughout. A great deal of time, money, and effort has gone into pre-attack reliability. Therefore, U.S. officials have confidence that U.S. ICBMs are very reliable. They probably are, but no one is 100 percent certain.

In a society where practically everything produced malfunctions on a regular basis due to an emphasis on meeting numerical quotas at the expense of quality, a Soviet warplanner would be remiss not to regard high Soviet ICBM reliability ratings with a slightly jaundiced eye. Indeed any military man who has experienced combat or who is a student of combat must retain a healthy respect for how quickly plans conceived far removed from the battle have to be revised under operational conditions. ICBM attacks may not be exceptions to this rule. As two analysts of this phenomenon have observed:

> Missiles...do not exist in the orderly universe of the strategic theologians but in the actual world of contract mismanagement, faulty parts, slipshod maintenance, bureaucratic cover-up, and the accidents that have afflicted military equipment since the world's first bow string got wet in the rain.[19]

This is not to argue against ICBMs as a weapon system. They are the most accurate, long-range weapon system in either the U.S. or Soviet arsenals. Their accuracy far outstrips that of the SLBM and their speed gives them superior penetration against defenses when compared to strategic bombers. But even the ICBM -- the maximum weapon available -- has its limits and there remains a margin of uncertainty regarding its level of wartime reliability.

Not only can missiles blow up on the pad or go awry after launch, but a full scale countersilo attack would require the coordinated cooperation of several hundred people whose behavior cannot be reliably

predicted under circumstances where they believe that
they and their families are about to die and where
their actions could lead to the deaths of millions of
others at the target end of their ICBMs. A Soviet
warplanner would have to take into account the
reliability of the human beings in the Soviet missile
firing loop. This could also be an American problem.
Journalist Jack Anderson has written that "there is
evidence that during one false alarm, a number of U.S.
crewmen ultimately charged with launching Minuteman
ICBMs did not follow official checklist procedures
during the six minutes when the computer indicated that
the United States was under attack."[20] Whether or
not this story is completely accurate or not, the fact
remains that no warplanner knows how well or poorly
people in charge of the attack (or counterattack) will
perform until the balloon goes up. A technically
perfect ICBM could be negated by a "malfunction" of the
launch team.

Other Soviet Weapons Uncertainties

The Soviet Marshalls and Generals assigned the job
of planning a future attack have to contend with many
other uncertainties regarding weapons performance.
Will the full Soviet ICBM force be generated and available at the time of launch? Would not generation to
alert status of Soviet bombers and ballistic missile
submarines, if not ICBMs, give the United States strategic warning of the impending attack and time to preempt or take complicating countermeasures?
One uncertainty might be the yield of the nuclear
warheads on the reentry vehicles. The uncertainty in
the average yield of a given kind of nuclear warhead
might be 10 percent or more.[21] Fuzing failures
could lead to detonation at less than the optimal
height of burst above the U.S. ICBM silos. A hitch in
the launch rate of Soviet ICBMs could lead to a poorly
timed attack that increased fratricide effects or which
allowed too much time for U.S. retaliatory launches
while the attack was underway.
The technical unpredictability and uncertainty of
nuclear war was summarized by former Secretary of State
Henry A. Kissinger in a speech several years ago. He
observed that:

No nuclear weapon has ever been used in modern
wartime conditions or against an opponent possessing means of retaliation. Indeed, neither side has
even tested the launching of more than a few
missiles at a time; neither side has ever fired
them in a North-South direction as they would have
to do in wartime. Yet initiation of an all-out
surprise attack would depend on substantial confidence that thousands of reentry vehicles launched

in carefully coordinated attacks -- from land,
sea, and air -- would knock out all their targets
thousands of miles away with a timing and reliability exactly as predicted before the other side
launches any forces to preempt or retaliate and
with such effectiveness that retaliation would not
produce unacceptable damage. Any miscalculation
or technical failure would mean national catastrophe. Assertions that one side is "ahead" by
the margins now under discussion pale in significance when an attack would depend on decisions
based on such massive uncertainties and
risks.[22]

Soviet Uncertainties About Fratricide and Other Local Effects

Perhaps one of the greatest uncertainties in
calculating the effects of a first strike by one
missile force against the ICBM force of another is
how to deal with fratricide effects.

Fratricide

Fratricide in the Biblical sense means the killing
of brother by a brother such as when Cain killed Abel.
Fratricide in the parlance of the nuclear strategist
is where the explosion from one nuclear warhead spoils
the attack of another warhead aimed at the same or
nearby enemy targets.

It is not certain what number of warheads can be
exploded on a given target within a short time before
subsequent attacking reentry vehicles are destroyed or
blown off course by the previous exploding warheads and
resulting winds, blast waves, thermal effects, effects
of EMP, IEMP, TREE, X-Rays, and neutron radiation, as
well as the damaging effects of dirt, rock, ice, and
water that are drawn miles up into the stem of the
nuclear mushroom cloud. The impact of fratricide is
likely to be greatest and least measurable when an
entire target complex, such as an ICBM field, is undergoing attack at the same time.

During the first wave of an attack there may be a
brief interval, measured in a few minutes or seconds,
during which fratricidal interference would be low
because the cloud stem with its debris had not yet had
time to propagate sufficiently to blanket the area. An
attacker might successfully explode a second or third
weapon on the target in this window of time before
fratricidal effects could blunt the attack, but whether
or not such precise time-on-target was achievable must
remain an uncertainty in the minds of Soviet planners
until after the event.

There are a number of ways to minimize or negate
possible fratricide effects that an intelligent

attacker may adopt. First, Soviet warplanners no doubt realize that air bursts cause much less debris to be drawn up into the atmosphere than ground burst weapon that excavates a large crater. Therefore, one strategy to avoid fratricide is to explode one or more air bursts against an enemy ICBM silo before finishing the job with a ground burst weapon. As much as a fifteen second window of time can exist between an air burst weapon and the subsequent ground burst, enough time for a sophisticated attacker with a modern ICBM force with good time-on-target control such as possessed by the Soviet Union or United States.

Second, the Soviet warplanner might avoid fratricide effects preventing attacks against adjacent targets by azimuthal selectivity. That is, they might plan to fire weapons launched from different missile fields in the Soviet Union against targets adjacent to each other in a single missile complex in the United States. In this way they could plot attack azimuths so that arriving weapons (coming in from different angles) could bypass the fireballs, mushroom clouds, and radiation effects created by earlier detonations in the target area. Azimuthal selectivity can lower the fratricide uncertainty involved in predicting the outcome of attacks on targets located in close proximity to each other.

Thus, an attacker, while respecting the disturbing effects of fratricide can, nevertheless, probably avoid most of the problem. Some nuclear effects experts working for the Department of Defense believe that attacks of 3 on 1 or 4 on 1 (RVs per silo targeted) can be achieved despite potential fratricide problems if some of these measures are taken.

One future counter to Soviet air burst weapons might be the superhardening of U.S. silos. Such a move could create problems for Soviet warplanners since overpressure from airbursts might not be sufficient for superhard silo destruction. On the other hand, if the silo is in the crater dug by a ground burst weapon (surface burst or an earth penetrator) the silo-launcher could not survive to function as intended.

Should the Soviet warplanners adopt a rollback attack, rather than a spike attack, other uncertainties would prevail. The adoption of rollback attack tactics raises uncertainties for Soviet warplanners because it might take additional time to complete any "South-North walk" through each U.S. ICBM field.

A rollback attack would help avoid fratricide effects, but would also give the U.S. leadership time to launch the as-yet-untouched portion of the U.S. ICBMs out from under the attack. It is for this reason that Soviet military officials might plan instead for a spike attack where at least one Soviet weapon would be

programmed to airburst simultaneously over every U.S. ICBM silo.
In such an attack some believe it is possible that a second, third, and perhaps even a fourth Soviet warhead could be delivered on each target a few moments later and still avoid the nuclear radiation, fireball, shock waves, nuclear winds, and nuclear cloud from the first explosion. Such a spike attack might require that the second, third, or fourth weapons detonated on targets be launched from separate launch sites in the Soviet Union and travel along vectors that carried them through the unperturbed spaces between nuclear eruptions.
Fratricide calculations are all theoretical and are not based on any test data of multiple shots tried in closely timed sequence. Therefore, a degree of uncertainty about the importance of fratricide still exists among U.S. experts, and presumably, among Soviet warplanners as well.

Air Density and Winds
Soviet military planners also would have to consider several other local climatic and geological factors that would be difficult to measure in advance when computing the possible outcome of a countersilo attack.
First, target-end air density varies with the season, weather and wind velocities. Air density is even sensitive to solar flares or sun spots. Air density, in turn, affects the amount of friction and drag that influences a reentry vehicle as it descends through the atmosphere toward the target below. This, in turn affects the accuracy of the RV and its lethality vis-à-vis the target.
Most of the time, under ordinary conditions, variations in air density and wind have little effect on accuracy. However, in extreme circumstances such as the presence of jet stream winds, significant errors could result. Soviet reentry vehicles that have to descend through high altitude (30,000 to 40,000 feet) jet stream winds of 180 knots might drift off target by 1,065 feet.[23] Such a wind-induced drift might mean the difference between a Soviet weapon killing or failing to destroy a hardened U.S. ICBM silo.
Air density and wind effects were considered much more significant sources of missile reentry inaccuracy when reentry vehicles were designed to come into targets at very low angles close to the earth's surface and through quite a bit of extra atmosphere. However, as one U.S. Air Force General notes:

with the advent of high ballistic coefficient reentry vehicles, this effect has become so minor that corrections are no longer made for climatology.[24]

Because of their irregularity, Air Force calculations ignore jet stream winds and assume normal wind conditions above the target.

Geology

Less important, local geology under U.S. ICBM silos, how deep the water table and bed rock is found, somewhat determines the degree of "bounce" of the nuclear shock wave against the silos and the shearing forces directed on its walls and headworks. ICBM silos placed in dry hard rock exhibit a different resilience to blast effects than those located in wet soft rock, dry soft rock, dry soil, or wet soil. Local geological factors effect the vulnerability or hardness of the targeted forces. To the degree that these factors are not precisely known to Soviet targeteers, a small degree of uncertainty again enters the calculations.

U.S. Forces, NCA and C^{3}I: Vulnerability and Uncertainty

Silo and Missile Hardness

Uncertainty about U.S. silo hardness cuts both ways. The Soviets cannot be absolutely certain that U.S. silos are not hardened above their advertised durability. U.S. ICBM silos could in some indeterminate number of cases withstand considerably more overpressure than what is presently believed to be their capacity.

The Limited Test Ban Treaty of 1963 banned all atmospheric nuclear tests by the United States, United Kingdom, and the Soviet Union. Prior to 1963 no one thought that missile accuracy would improve so greatly that it could eventually pose a serious threat to rival ICBMs in their silo-launchers a continent away. Therefore, neither the United States nor the Soviet Union have conducted any testing of nuclear airbursts against hardened silos.

U.S. and Soviet silos have since been tested against the shock waves of conventional ordnance exploded overhead and underground nuclear tests detonated nearby. U.S. scientists and officials believe they know within a narrow margin of uncertainty what "hardness" to assign a given silo.

Recent silo test results raise the possibility that future ICBM silos and associated missiles might one day be hardened to more than 25 times current levels. This quantum leap in silo durability might more than keep pace with projected Soviet advances in

missile accuracy. Superhardening U.S. silos could add
considerable uncertainty to Soviet calculations.
If the U.S. ICBM silo escapes destruction from
blast overpressure or the adverse effects from
cratering, this does not necessarily mean that the
Soviet attack was unsuccessful. The vibration and
shock of the attack might cause the missile to malfunction subsequently upon launch even if the silo itself
escaped serious destruction or displacement. This is a
case where uncertainty works to the disadvantage of the
defender. Soviet warplanners might hope that attacks
that were unsuccessful against U.S. missile silos might
nevertheless be successful against the missiles within
those silos. U.S. designers of ICBM shock isolation
systems believe that ICBMs can be protected against
blast waves to the degree that they can survive any
explosion their silos can withstand.
Another spectre facing a prudent warplanner is the
chance that one's own intelligence is faulty and has
failed to discover a clandestine cache of strategic
nuclear weapons or has failed to identify where spare
ICBMs are held in reserve. Such extra missiles could
be fired from launchers at presurveyed hard points and
would not necessarily need to be based in silos. This,
no doubt, is mainly a U.S. targeting problem, but the
uncertainty may yet remain in Soviet thinking about
whether the U.S. is hiding retaliatory power.

NCA and C³I Targeting Uncertainties

One of the most tempting strategies to Soviet
nuclear warplanners might be a precursor decapitation
strike against U.S. leadership and C³I targets
followed immediately by a massive strike against U.S.
strategic and theater nuclear forces.
It is part of stated Soviet military doctrine that
they would favor nuclear decapitation attacks and make
U.S. command, control, and communications assets targets of the highest priority.
If a major weakness exists in U.S. deterrence
strategy, it is the vulnerability of the President and
the other NCA designates to a Soviet decapitation
attack. Not only is the leadership endangered but so
too are the strategic command, control, and communications links between the U.S. leadership and U.S. strategic forces.
If NCA vulnerability is the glass jaw of the U.S.
deterrent posture, then C³I vulnerability is its
potential Achilles heel. It may matter little if the
U.S. National Command Authority survives if it cannot
communicate orders to U.S. forces due to destruction of
the communications links between them. Former SAC
Commander Curtis LeMay once said, "without communications, all I command is my desk."25 The NCA could

be in the same situation in the opening minutes and
hours of a nuclear war.
However, the Soviet warplanner could never know
for sure whether or not U.S. forces would execute the
U.S. Single Integrated Operating Plan (SIOP) if the
National Command Authority was killed or temporarily
silenced by an attack.
Soviet warplanners would have to consider the
possibility that the President would have predelegated
his authority to U.S. commanders to use U.S. nuclear
weapons in certain emergency conditions.
Soviet attack planners would have to consider
whether or not:

> The (U.S.) presidential center served as a safety
> catch, holding back the multiple triggers embodied
> in the primary commands, corresponding to the
> unified and specified commanders (SAC, CINCPAC,
> CINCLANT, CINCEUR). A Soviet attack on the
> presidential command center would then be an
> attack on the safety catch of the entire command
> structure, and the Soviets would be destroying the
> one mechanism holding back all-out retalia-
> tion.[26]

It appears that the Soviet warplanner's only
possible strategy for possibly paralyzing a U.S.
nuclear weapons retaliation is a nuclear decapitation
strike. However, such a Soviet strategy could backfire
badly, virtually guaranteeing an unrestrained U.S.
nuclear response if Presidential launch-on-warning or
launch-under-attack orders were predelegated to U.S.
commanders.

Conclusions: Uncertainty and Deterrence

Even if the Soviet Union acquired the theoretical
ability to destroy every U.S. ICBM in its silo, the
United States still ought to be able to deter a Soviet
attack by virtue of the great retaliatory capability
that still resides in its SAC bomber force and its
fleet ballistic missile submarines.
The thrust of this analysis is that the Soviet
warplanner would have many uncertainties in estimating
the outcome of any attack on U.S. ICBMs. But beyond
this analysis, it is important to keep in mind that
U.S. ICBM vulnerability to Soviet attack is not the
same as triad vulnerability.
The triad of forces is useful for long term
stability. If ICBMs are vulnerable, the SAC alert
bomber force and the Poseidon/Trident SSBN force can
compensate for the deficiency and provide deterrence
"cover" until ICBM survivability is upgraded.

The argument that uncertainties ought to help Soviet leaders decide on the side of caution also presumes rational Soviet decision-making during the stress of an acute crisis. Some of the well documented symptoms of tension overload on individuals include panic, stereotyping, simplification of issues, immobilization or impulsive behavior, and a reduced ability to discriminate between fine points attached to different decision paths. The stress inherent in any superpower confrontation, given the stakes involved, might be enough to unhinge a normally unflappable policy maker and might make subtle uncertainty arguments difficult to make or understand.

Up to now, with the exception of the 1962 Cuban Missile Crisis, the Soviets have tended to act very cautiously, intervening in "cases where they have supreme interests at stake, a high probability of U.S. noninvolvement, and a comfortable prospect of success with moderate investment of military capital."[27] The Soviets appear to have adopted a risk minimizing strategy vis-à-vis U.S.-Soviet confrontations.[28]

Any factor that creates uncertainty in the Soviet warplanner's mind as to the potential effectiveness of a contemplated attack against U.S. ICBMs in their silos contributes to deterrence, peace, and stability. Certain factors, such as the U.S. strategy in responding to strategic and tactical warning of such a Soviet attack are beyond the Soviet warplanner's power to affect. The possibility of a U.S. launch-on-warning and launch-under-attack strategy should give any Soviet leader pause who is weighing the option of striking first.

Even if the United States were to adopt a "ride-out" strategy in the event of a Soviet countersilo attack, the Soviet attack planning staff could never be sure of achieving anything like complete destruction of the U.S. ICBM force to say nothing about their chances of executing the far more difficult task of simultaneously destroying the U.S. fleet ballistic missile submarine force or the alert SAC bomber force. The uncertainties are so great when added together that the Soviets might destroy nearly all U.S. ICBMS or they might fail to destroy more than a handful.

Would a rational Soviet leader who understood these uncertainties risk such a first strike? Even 20 surviving Peacekeeper missiles would have the potential to destroy the 200 most important bunkers and silos in the U.S.S.R. and might put at risk the top CPSU leadership.

No prudent Soviet warplanner who understood the wide spectrum of possibilities would risk attacking the Minuteman or Peacekeeper force in view of the great uncertainties in predicting the outcome of such an

attack and the massive penalties that would be visited
on Soviet leaders and the Soviet Union if they guess
wrong.

Although the vulnerability of U.S. silo-based
ICBMs has been oversold somewhat in some quarters, from
the perspective of the conservative U.S. defense
analyst this is still a legitimate worry in the face of
the Soviet SS-18 and SS-19 countersilo threat. Nevertheless, it is instructive to look at the strategic
world through the other end of the telescope as well.
The conservative Soviet warplanner will see numerous
very important uncertainties of outcome when he
attempts to calculate the results of potential Soviet
first strikes against U.S. ICBMs in silos.[29] The
greater his pessimism and the deeper his uncertainty
about attaining a successful military outcome from any
first strike attack the safer the United States will
remain.

NOTES

1. Report of the President's Commission on Strategic
    Forces (Washington, D.C.: April 6, 1983), p. 17.

2. Harold Brown, Secretary of Defense, Speech to the
    Commonwealth Club of San Francisco as reported in
    "Brown Says Launch-On-Warning Cannot Be Ruled
    Out," Defense/Space Daily, June 27, 1978, p. 286.

3. See Kosta Tsipis, "Precision and Accuracy," Arms
    Control Today, Vol II, No. 5, May 1981, p. 3.
    Once a missile has been fully tested, unlike a
    rifle, it becomes a "closed loop" system that
    takes some self-correcting measures during flight
    to eliminate accuracy errors. Therefore, even on
    the first shot at an unfamiliar target, ICBM
    accuracy is likely to be better given the range of
    flight than rifle accuracy on the first test
    shot.

4. James Fallows, National Defense (New York: Random
    House, 1981), p. 153.

5. Ibid., p. 153.

6. Ibid., p. 153.

7. Andrew and Alexander Cockburn, "View From the
    Fourth Estate: The Myth of Missile Accuracy,"
    Parameters, Journal of the U.S. Army War College,
    Vol VI, No. 2, p. 87. The original source is D.G.

Hoag, "Ballistic Missile Guidance" in Impact of New Technologies on the Arms Race (Cambridge, MA: MIT Press, 1971).

8. Paul S. Mann, "Panel Reexamines ICBM Vulnerability" Aviation Week and Space Technology, Vol. 115, No. 2, p. 141.

9. Ibid., p. 141.

10. Gen. Robert T. Marsh, USAF, "Strategic Missiles Debated: Missile Accuracy -- We Do Know!" Strategic Review, Spring, 1982, p. 36.

11. Ibid., p. 36.

12. This is the consensus of USAF scientists who are aware of these sources of CEP error. Continental drift, for example, might throw accuracy off by a centimeter or two. Gravitational effects of the sun or moon might account for five feet of inaccuracy. Solar wind might change CEP by an inch. These effects are small enough to be neglected.

13. Again, USAF engineers believe they understand the implications of range extensions on bias and CEP with a high degree of confidence. So long as the RV explodes within a lethal range of the target, and ICBM error budgets correctly predict that the impact point will be within that range, such inaccuracies are tolerable from a military effi* ency perspective.

14. See Paul S. Mann, "Panel Reexamines ICBM Vulnerability," Aviation Week and Space Technology, July 13, 1981, p. 148. See also: J. Edward Anderson, Op. cit., p. 39.

15. Ibid., p. 148.

16. Cockburn, Op. cit., p. 85.

17. Matthew Bunn and Kosta Tsipis, "Ballistic Missile Guidance and Technical Uncertainties of Countersilo Attacks," Report No. 9, Program in Science and Technology for International Security, Department of Physics, MIT, August 1983, p. 83.

18. Ibid., p. 83-84.

19. Cockburn, Op. cit., p. 85.

134

20. Jack Anderson, "Can We Have A Nuclear Accident?" Parade Magazine, August 10, 1983, p. 10. Anderson reported that "at McConnell (AFB) only four of the 17 crews followed the prescribed procedure. At Grand Forks, N.D., two of the 13 crews didn't insert their keys until first talking it over. Then it would have been too late."

21. Bunn and Tsipis, "The Uncertainties of a Preemptive Nuclear Attack," Scientific American, Vol. 249, No. 5 (November 1983), p. 43.

22. Henry A. Kissinger, Secretary of State, Speech before the World Affairs Council and Southern Methodist Unversity in Dallas, Texas, on March 22, 1976. (Reprinted by the Office of Public Affairs, Dept. of State).

23. Cockburn, Op.cit., p. 89. See Footnote 7.

24. March, Op. cit., p. 36.

25. Quoted by Gen. Kelly H. Burke (USAF, ret.), "Strategic C^3: Why It Counts," Armed Forces Journal International, February 1983.

26. Paul Bracken, The Command and Control of Nuclear Forces (New Haven, CT: Yale University Press, 1983), p. 202.

27. Benjamin S. Lambeth, "Uncertainties for the Soviet War Planner," International Security, Vol. 7, No. 3, Winter 1982/1983, p. 143.

28. Cockburn, Op. cit., p. 85.

29. Ibid. See fn. 4. Lambeth cites Hannes Adomeit "Soviet Risk-Taking and Crisis Behavior: From Confrontation to Coexistence," Adelphi Papers, No. 101, (London: IISS, 1973).

8
The Soviet Union and Modernization of the U.S. ICBM Force

Dan L. Strode

Introduction

Modernization of the U.S. ICBM force has been an essential element of the defense programs of the past three administrations. The Ford Administration initiated the Peacekeeper program to provide a counterweight to Soviet deployment of a fourth generation of ICBMs and to enable U.S. forces to carry out the increasingly rigorous missions required by U.S. deterrence policy. Beginning in 1974-1975, the Soviet Union began deploying three and possibly four new ICBMs, the SS-16, SS-17, SS-18, and SS-19. The latter three missiles incorporated MIRV technology, providing the Soviets with this capability sooner than anticipated. Moreover, these ICBMs had greater accuracy and throwweight than the systems they replaced. This combination of accuracy and throwweight, when combined with MIRV technology, promised to provide the Soviet Union with an advantage in the ability to destroy very hard targets, such as ICBM silos and command bunkers. Moreover, the new Soviet missiles were housed in improved silos, decreasing their vulnerability to attack by U.S. missiles. The Peacekeeper was originally designed to help offset this potential asymmetry in countersilo capabilities.

Changes in U.S. strategic doctrine originating during the Nixon Administration and coming to fruition in the so-called Schlesinger Doctrine provided an additional rationale for the Peacekeeper. The Schlesinger Doctrine required U.S. forces to be able to execute a variety of limited nuclear options, varying in scale and purpose. Accuracy, retargetability, and redundancy of command and control made a modernized ICBM force the logical mainstay of the new strategy.

The Carter Administration further developed the themes of the Schlesinger Doctrine in PD 59. According to administration spokesmen, the new doctrine was specifically designed to "take into account and assist in shaping Soviet perspectives."[1] Secretary of Defense Harold Brown identified four Soviet beliefs as central to the new U.S. strategy: (1) the possibility that nuclear exchanges might be prolonged; (2) that military forces are the prime targets; (3) that the Soviet regime values its own survival above all else; and (4) that some elements within the Soviet leadership believe in the possibility of victory in a nuclear war.[2]

PD 59 reportedly stressed targeting flexibililty and ad hoc planning. The ability to threaten Soviet military forces and political control centers, not just general economic targets, was deemed important to deterrence. To permit the flexible and discriminate use of U.S. forces, PD 59 required that C^3I be resilient. Some nuclear forces were also to serve as a secure reserve force. The goal of force employment would not be to "win" a nuclear war with the U.S.S.R., but to deny the Soviets victory on their own terms and to provide leverage for a negotiated termination of the fighting.[3] These new capabilities would allow the U.S. to deter the Soviets from any level of attack, particularly from a "limited" strike against the United States.[4] The Reagan Administration reportedly accepted the new strategy and sought to speed its implementation by a more comprehensive acquisition policy.

For both the Carter and Reagan Administrations, a modernized ICBM force has been essential to implementation of the new strategy. The Carter Administration hoped to provide the combination of survivability, accuracy, and responsiveness to control necessary to support the "countervailing strategy." The Reagan Administration is seeking to obtain the same basic capabilities through two separate systems, the Peacekeeper deployed in silos at Warren AFB and a mobile small ICBM (SICBM).

Given the centrality of ICBM modernization to the overall revitalization of the U.S. strategic posture under the last three administrations, it is not surprising that the Soviet Union has made substantial efforts to retard the development of Peacekeeper ICBMs through political means and to offset their potential effect on Soviet military capabilities. To understand these Soviet efforts, it is necessary first to analyze Soviet evaluations of the relationship between Peacekeeper, the SICBM, and U.S. nuclear policy.

ICBM Modernization in U.S. Strategy -- the Soviet Perspective

Soviet analysts of American defense policy do not treat any single class of weapon in isolation; they place systems within the framework of military strategy and total force posture. Improvements in the U.S. ICBM force in this view are integrally related to changes in American nuclear policy. General Mil'shtein has argued that:

the appearance of highly accurate nuclear missile weapons -- and in considerable quantity -- has led the United States to undertake a review of a number of fundamental military-strategic concepts in a highly dangerous direction, especially those which examine the possible consequences of a nuclear war and schemes for utilizing nuclear missile weapons in such a war.[5]

Soviet critics of American defense policy attacked both the Schlesinger Doctrine and PD 59 for attempting to establish a useable margin of nuclear advantage with which to blackmail the Soviet Union.[6] General R. Simonian argued that the Peacekeeper ICBM was the primary means by which the United States hoped to obtain this advantage.[7]

During the late 1970s and early 1980s, Soviet analysts could not agree whether the new American strategy was a real departure from the past. For example, Lev Semeiko insisted on the continuity of PD 59 with American post-war nuclear strategy:

The New York Times calls the directive a "new U.S. nuclear strategy." One wonders what is new about it. After all, back in the first half of the sixties, the "counterforce" concept was elaborated in the United States -- the concept of inflicting nuclear strikes on military targets in the attempt to deprive the enemy of the opportunity to resist.[8]

According to this line of Soviet reasoning, there is nothing new in U.S. strategy since the United States has always planned a first strike against the U.S.S.R. Semeiko argued:

During the postwar period there have been many changes of administration in the United States and U.S. military doctrines have frequently changed with these. However, the desire has remained unchanged to create and, at a suitable moment, to use a strategic forces capability to

inflict a preemptive, disarming nuclear strike.
The attempts to hide this desire have been
equally unvaried. Officially the talk has been
and still is of preparing for a "counter"
(second) strike. But these are just words
designed to hide the Pentagon's real military-
strategic aims.[9]

Other Soviet analysts argued that until PD 59,
"assured destruction" remained "basic to U.S. military
doctrine."[10] The new strategy was "a _potential_ first
strike concept," and it demonstrated that those who
favored developing the ability to conduct a nuclear
war, as opposed to those who merely sought to deter
war, were gaining the upper hand.[11]

President Reagan's success in sustaining the
basic elements of his strategic forces modernization
program during the first half of his term and his
administration's adoption and extension of PD 59
helped create a new consensus within the Soviet leadership about the prospects of altering American
policy.[12] Today Soviet spokesmen accuse the United
States of making active preparations to launch a surprise nuclear attack, and they point to the Administration's ICBM modernization program as evidence.[13]

The Soviets have also expressed their displeasure
with the United States' announced intention to give
greater emphasis to counter-political targeting. As
the formulators of PD 59 assumed, the political power
of the regime is its top priority, and counter-
political targeting has aroused the concern of Soviet
analysts. Former Secretary of Defense Harold Brown
argued that favorable war termination and escalation
control require that high value Soviet political,
economic, and military targets not be destroyed in the
initial exchange, but be held at risk.[14] The Soviets,
however, interpreted counter-political targeting
differently. Lev Semeiko maintained:

Now the list of targets includes political and
military control centers, which is not a mere
arithmetic increase in the number of targets
(roughly 50 percent, according to some American
data), but an obvious attempt to win a victory in
a nuclear war, the possibility of which the American leaders themselves verbally reject in view of
the catastropic consequences it would have. It
is difficult to draw any other conclusion, for
the destruction of the entire complex of strategically important facilities can, according to the
Pentagon's calculations, paralyze the enemy and
force him to capitulate, rather than merely
impressing him by individual strikes...."[15]

It must be emphasized that there is a definite propaganda aspect to Soviet public interpretations of all American strategies as methods of warfighting. It should be recalled that Soviet authors also described Mutual Assured Destruction as a plan to win a nuclear war by destroying the enemy's vital economic-industrial centers. Yet there is a deeper reason for the Soviet reading (or misreading) of U.S. strategic policy. The Soviets translate U.S. concepts into the thought framework established by their own tradition of strategic thinking, which has a decidedly opera-tional focus. For example, Major General Simonian described the complex discussions of deterrence, extended deterrence, and escalation control which went into the formulation of PD 59 as "questions connected with preparation for war and waging of such a war..."[16] In the Soviet tradition of strategic thought, the ability to destroy the enemy's military forces and political organization with a sudden attack is essential for obtaining full victory while limiting damage to the Soviet Union. Soviet analysts have a natural tendency to read American strategic thought in this light. Thus when U.S. strategists argue for the need to enhance deterrence by obtaining the ability to threaten Soviet political and military targets, many Soviet analysts suspect that in realilty the United States is seeking a preemptive strike capability. The Soviets do their own mirror imaging when analyzing U.S. programs and strategy.

Soviet analysts discount U.S. claims that either the Peacekeeper or the SICBM are designed for second-strike, counterforce missions and controlled response, and they are quick to point out that counterforce cap-abilities are most effective when used first. General Simonian pointed out, "Indeed a power which sets itself the aim of destroying the 'potential' enemy's military facilities must be first to deliver a strike because otherwise its nuclear charges will land on empty missile silos and airfields."[17]

One important reason why the Soviets will not accept declared U.S. purposes and rationales for ICBM modernization is that they publicly reject the limited war hypotheses upon which they rest. The Soviets have rejected such notions in their declared statements since they were first put forward by Secretary of Defense McNamara in 1962. The authors of __Military Strategy__ wrote:

As a matter of fact, how can anyone be "con-vinced" of the necessity to adhere to the 'new rules' that nuclear strikes should be launched only against military objectives and not against cities, when the majority of such objectives are located in large or smaller cities and populated places?[18]

Minister of Defense Ustinov has derided the notion that nuclear war can be conducted "in a gentlemanly fashion,"[19] and Chief of the General Staff Ogarkov has argued,"..in practice the idea of restraining a nuclear war within some limited framework is utopian."[20]

In the Soviet literature, there are variations over time and between sources on the feasibility of restraining a nuclear war to Europe and on the possibility of a large scale purely conventional war. Moreover, Soviet statements on limited war are influenced by propaganda concerns. However, the majority of serious Soviet military literature does not advocate structuring large scale intercontinental nuclear exchanges to enhance escalation control. Writing in Kommunist, the leading theoretical journal of the Communist Party of the Soviet Union (CPSU), Alexei Arbatov expressed the Soviet belief that current U.S. doctrine for counterforce warfare is irrational:

If Washington planners mean a coordinated missiles strike against the entire complex of ground-based strategic targets, that would mean, in effect, dropping several thousand nuclear warheads on the territory of another country. In that case, the aggressor can be sure that the reply to such a 'selective' strike would be a full-scale devastating retaliation possibly without waiting for the 'counterforce' attack to hit its targets...[21]

In summary, the Soviets indicate that they see U.S. ICBM modernization as an integral part of a new strategy whose goal it is to paralyze the U.S.S.R.'s political and military command structure with a surprise attack and then destroy Soviet nuclear forces with counterforce strikes. The Soviets seem to find the notion of limited nuclear war associated with the new strategy difficult to accept and suspect that it may be a cloak behind which to wage full scale war.

Soviet Strategy

A fuller understanding of Soviet attitudes toward U.S. ICBM modernization requires an examination of the U.S.S.R.'s military policy, of how different weapons systems fit into its overall strategic calculations. Soviet strategy also provides a sound basis for judging the validity of the rationale for deployment of

Peacekeeper and the SICBM. This survey of Soviet strategic thinking does not pretend to be complete,
but focuses instead on those facets with the greatest
relevance for ICBM modernization -- targeting priorities and the duration of war.

Soviet Targeting Priorities

The countervailing theory of deterrence adopted
by the Carter and Reagan Administrations focused
primarily upon developing a retaliatory threat capable
of holding the highest values of the Soviet leadership
at risk -- their lives, their political and military
control, their war industry. PD 59 and developments
under the Reagan Adminstration also seek to enhance
deterrence of war by providing proportionate U.S.
responses to Soviet aggression -- even if that aggression is at the strategic nuclear levels.

Some U.S. analysts fear that, if the Soviet Union
launched a limited attack against ICBMs and other
nuclear forces, the President would face an unacceptable choice. He would either have to acquiesce to a
Soviet imposed resolution of the crisis or to initiate
a suicidal full scale nuclear exchange.[22] Modernization of the U.S. ICBM force has been designed to
thwart such an attack and to broaden the range of
possible responses.

For example, the SICBM may use mobility to survive the initial attack and yet should still be accurate enough to threaten remaining Soviet nuclear
forces and key political targets. The joint deployment of Peacekeeper and small ICBMs could stress
Soviet war plans because a force optimized against
silo-based ICBMs would probably not be optimized
versus mobile systems. Presumably, in response to
U.S. restraint in avoiding political targets and
population centers, the Soviet Union will refrain from
destroying major urban areas.

However, it is not clear that the fear of a
"limited" intercontinental attack by the U.S.S.R.
adequately takes into account Soviet targeting policy
as depicted in their strategic writings. Of course,
we do not know for sure what actual Soviet targeting
plans are. Yet, to the extent that war plans are
reflected in doctrinal writings, the following
considerations should hold.

American strategic thought has long depicted targeting priorities as falling into two categories --
counterforce and countervalue. However, Soviet strategic writings have never used the term "counterforce"
or "countervalue" to describe their own targeting policies. Instead, they speak of the need rationally to
allocate weapons between "passive" and "active"
targets.[23] Active targets are, primarily, nuclear

weapons and their means of delivery, while passive targets include political-administrative centers and economic targets. All three editions of the Soviet volume __Military Strategy__ listed nuclear weapons, their delivery systems, the war economy, government and military leaders, communications, and reserves as legitimate targets.[24] This essential targeting list is repeated with the addition of "groups of forces" in the 1982 work __Military Technical Progress and the Armed Forces of the U.S.S.R.__[25] For our purposes here, it is of interest that the Soviets intend to attack the war economy and "political-administrative" centers. That is, major urban areas over and above those close to military installations probably are targeted. Colonel Shirokov has stated:

> For the achievement of victory in a present-day nuclear war, if it is unleashed by the imper- ialists, not only the enemy's armed forces, but also the sources of his military power, the important economic centers and also points of military and state control as well as areas where different branches of the armed forces are based will be subjects to simultaneous destruction.[26]

Given the high readiness rates and quick reaction capabilities of modern nuclear forces, particularly ICBMs, the Soviets point out that it is difficult to destroy nuclear forces on the ground. Hence attacks on command and control are increasing in importance.[27] Shirokov praised the potential of such attacks:

> Under conditions of nuclear war, the system for controlling forces and weapons, especially stra- tegic weapons, acquires exceptionally great significance. A disruption of control over a country and its troops in a theater of military operations can seriously effect the course of events and in difficult circumstances, can even lead to defeat in a war.[28]

The emphasis on counter-control targeting will broaden the scope of any "limited" Soviet attack. Moreover, it may be difficult to conduct war flexibly with escalation control and early war termination if political and military control centers have been attacked.

Since the Soviets insist that important sectors of the enemy's war economy must be attacked at the outset of a conflict, some American cities probably will not be spared in the initial phase of combat. However, according to Shirokov, attacks on cities are not indiscriminate:

> The objective is not to turn the large economic
> and industrial regions into a heap of ruins
> (although great destruction, apparently, is
> unavoidable), but to deliver strikes which will
> destroy strategic combat means, paralyze enemy
> military production making it incapable of
> satisfying the priority needs of the front and
> rear areas and sharply reduce the enemy capabil-
> ity to conduct strikes.[29]

Soviet analysts study the U.S. economy for key
links, weak points whose destruction affects the total
economy and, most importantly, the war effort. These
include energy plants, transportation networks, and
oil refineries.[30] This is not a Soviet targeting
policy designed deliberately to produce high casual-
ties. However, it is not one which avoids civilian
loses at the price of sparing important military and
economic targets. The inclusion of industrial targets
further broadens the scope of any "limited" Soviet
attack.
   Population centers near military targets,
important links in both the political and military
chain of command, and important economic areas
probably will all suffer in the first phase of the war.
The Soviets cannot be confident that this type of
attack will be treated by the United States as
anything less than full scale. Therefore they are
dubious of the possibility of a controlled, large
scale war.
   Of course, Soviet critiques of American limited
war theories are partly propagandistic. They seek to
undermine the attempts of Western governments to take
measures necessary to prepare for contingencies short
of all-out nuclear war. Sometimes in their writings,
for purely internal consumption, Soviet military
writers show more interest in limited war. However,
these instances usually presume a continental theater
conflict where firepower must be coordinated with the
action of friendly troops. There are very few cases
in which the Soviet hint at the possibility of limited
intercontinental war.
   While the Soviets may not be impressed by U.S.
limited war concepts, they must be impressed (and
deterred) by the potential capabilities of the Peacekeeper and the SICBM.
   In a general nuclear war, the primary goal of
Soviet strategy is to destroy the opponent's nuclear
systems as rapidly as possible with the surprise use
of mass nuclear strikes. The Soviets argue that the
prompt destruction of the opponent's nuclear forces

must produce a dramatic change in the "correlation of forces" in their favor. General Anureev, the Soviet Union's leading systems analyst, described the purpose of massed nuclear strikes:

> One of the most important features connected with the application of nuclear weapons is the possibility of a sharp change in the correlation of forces. Skillful planning of offensive operations and their successful realization have always led to a change in the correlation of forces to the advantage of the side which has prepared carefully for the operation. However, it was not so sharp, so spasmodic, as it could be during the use of nuclear weapons.
> A sharp change in the correlation of forces to one's own advantage can be achieved by means of the mass application of nuclear weapons with the simultaneous repulsing of a sudden attack by the air-space means of the enemy, and in the process the compulsory condition of the optimal distribution of nuclear weapons carriers against enemy targets.[31]

Soviet force structure reflects this preemptive counterforce philosophy. The Soviet Union has placed most of its strategic inventory into rockets -- about 70 percent of its total warheads on intercontinental systems. Accuracy, speed and relative invulnerability in flight make the U.S.S.R.'s ICBM force the prime mechanism for executing the strategy described above. The SS-18 Mod. 4 and the SS-19 Mod. 3 are sufficiently accurate to put current U.S. ICBM silos at risk. However, there would be considerable uncertainty about whether the Soviets could successfully execute a surprise attack against U.S. ICBMs in their silos (see Chapter 7). Soviet Military Power predicts that the Soviets will deploy additional follow-on ICBMs with even greater accuracy in the near future.[32] Given the centrality of the ICBM in the Soviet Union's strategy and force posture, modernization of the U.S. ICBM force should enhance Soviet respect for the overall capability of United States' deterrent.

Soviet Strategy and Protracted Nuclear War

Current U.S. strategy recognizes the need to prepare for a protracted nuclear conflict. In support of this, it is customary to cite Soviet doctrinal writings and military exercises which suggest that the U.S.S.R. takes protracted war seriously. But what precisely does the Soviet Union understand by protracted war and what implications does this have for the modernization of the U.S. ICBM force?

The task of comparing the protracted war concepts of the U.S. and the U.S.S.R is difficult since there is considerable vagueness and incompleteness in the formulations of both sides. Neither has specified a time frame for example. Neither has adequately devised a system for mobilizing the nation's resources in a nuclear environment.

The issue of protracted war has been the subject of contention within the Soviet leadership in the past. Khrushchev supported a short war emphasis while many of his military advisors argued that the country must be prepared for both long and short conflicts.[33] By 1963, the latter position was predominant and remains official policy today.[34] Moreover, since 1974, there has been a distinct shift of emphasis in Soviet military writings which indicates that the likelihood that a war would be protracted has been upgraded.[35]

It is important to note that the Soviets do not see short and prolonged war strategies as contradictory. Rather, they believe that preparation for both eventualities to be a prudent hedge against the uncertainties of war and against the possibility that the enemy cannot be finished quickly. As Marshal Ogarkov, Chief of the General Staff, argued in 1979, "...taking into account the huge military and economic potentials of the coalitions of belligerent states, it cannot be excluded that it [the war] might also be prolonged."[36] The Soviets thus see failure to prepare for a long war as a reckless underestimation of the potential enemy's strength. Analogies to Hitler's failed attempt to defeat the U.S.S.R. rapidly are often stressed in this regard.[37] In addition, a protracted war cannot be excluded because of the uncertainties surrounding nuclear war. Will it start at a tactical level and then escalate? Will it begin with a conventional phase? Will it start as a local war? The Soviet authors of Military Strategy wrote, "It is quite obvious that a new world war cannot be reduced to some single scheme in as much as the concrete circumstances may produce the most varied and sometimes unexpected situations."[38] The Soviet Government's main objective is to win quickly. Prolonged war is not a preferred option, but a possibility for which planners must allow.

The Soviets not only discuss protracted war, they have made preparations for such an eventuality. They have practiced reloading SS-18 launchers in silos for example. Moreover, the U.S.S.R. may have stockpiled ICBMs for possible use after the first days of a nuclear war.[39]

Soviet plans for intra-war mobilization and the use of strategic reserves (including reserves of strategic rockets) have a direct impact upon the U.S. need

for modern ICBMs with counterforce potential. The
Soviets probably would follow up their initial waves
of attacks with further strikes by strategic reserve
forces. If circumstances permit, the Soviets would
also begin further attacks with mobilized reserves.
Thus the Soviets could have echeloned missile attacks
against the continental United States using both
deployed and reserve ICBMs.

In order to prevent the Soviet Union from
bringing the full weight of its mobilized nuclear
potential to bear on the United States in a conflict
(and thereby to deter such an attack), U.S. forces
must have the capability: (1) to disrupt the initial
wave of a Soviet attack; (2) to destroy Soviet
missiles held in reserve either in silos or in
stockpiles; (3) to destroy Soviet silos so that they
cannot be reused; and (4) to disorganize the control
of Soviet missile troops so that they cannot reconsti-
tute their ICBM force. This requires forces able to
respond promptly and accurately -- both potential
attributes of the Peacekeeper and the SICBM.

Soviet Strategy: Has There Been a Change?

In the early 1980s statements by Soviet leaders
and commentators indicated that a change might be
underway in Soviet strategic thinking. Leonid
Brezhnev stated in _Pravda_, "I will add that only he
who has decided to commit suicide can start a nuclear
war in the hope of emerging victorious from it."[40]
Alexei Arbatov claimed in the authoritative journal
Kommunist that "counterforce superiority is losing its
real military-political significance because, with the
present balance of forces, it cannot signify a disarming strike capability."[41] The current General
Secretary Konstantin Chernenko added his voice to the
chorus, arguing that it was "criminal" to view nuclear
war as a "rational and all but 'legitimate' continuation of politics."[42]

At the same time, other Soviets continued to
invoke victory as the goal of war. For example, the
head of PVO (the air defense forces) argued, "This
high degree of skill enables our servicemen to wield
modern weapons and equipment more effectively with a
view to attaining victory over a powerful technically-
equipped opponent."[43] A leading author of the Military History Institute wrote "the study of Leninist
theses on the role of the military factor in attaining
victory has great significance for Soviet military
science."[44]

There are several possible explanations for this
divergence of views. First, certain Soviet spokesmen
may have tried to mislead the West while Soviet
officers continued to prepare for the possibility of

war in the same way as in the past. Different statements in this case would be the product of poor coordination between various elites. Second, another
interpretation might be that while Soviet political
leaders do not necessarily believe that a meaningful
victory can be achieved as a result of a nuclear war,
they may nevertheless assign victory as a goal that
their armed forces must be prepared to achieve should
war occur. Or, third, there may have been a real
difference of opinion over nuclear strategy. Given
the authoritativeness of the sources cited, the latter
case seems more likely.[45]

Since Brezhnev's death, the debate seems to have
cooled down with the traditional military view of
strategy still intact. For example, Colonel G. Lukava
argued in a January 1984 article in Communist of the
Armed Forces:

> The most important duty, the pivotal direction of
> the activities of commanders and political
> workers, of Party and Young Communist League
> organizations, of the entire military society are
> the instilling in the personnel of confidence in
> the superiority of Soviet military science over
> bourgeois military theory and the conscious aware-
> ness of the necessity precisely and unfailingly
> to implement the principles of the science of
> victory set forth in the military regulations.[46]

The Soviet military continues to receive funds
for perfecting counterforce ICBMs, for air defense,
and for civil defense. Recent strategic war games and
organizational changes within the Soviet military are
also consistent with the Soviet strategy described
here.[47] Given the reassertion of traditional doctrine
after Brezhnev's death, the direction of organizational changes and the continuity of force development, the Soviet strategy described above probably
remains relevant.

Soviet Responses to U.S. ICBM Modernization--Military

Minister of Defense Dimitrii Ustinov and other
military leaders have expressed nervousness over
potential United States counterforce capabilities (as
should be expected).[48] However, Soviet force modernization appears to be keeping pace or exceeding U.S.
force improvements.

Soviet strategic initiatives already in progress
can address the problem created by improved U.S. ICBMs
in several ways. The Soviets will seek improvements
in missile and warhead technology adequate to retain
and to improve their ability to attack U.S. ICBMs with

a preemptive strike. If U.S. silos are made harder,
Soviet ICBMs will have to become more accurate to
retain present damage expectancies. According to
Soviet Military Power, the new SS-X-24 and SS-X-25
ICBMs, plus improved follow-ons to the SS-18 and
SS-19, will have improved accuracy.[49] If the United
States deploys large numbers of SICBMs in a mobile
mode, the Soviets may need to fractionate their force
loadings on ICBMs, reducing yield and possibly
accuracy.[50] Thus the requirements of simultaneously
being able to attack Peacekeeper ICBMs in very hard
silos and also dispersed mobile SICBMs may drive the
Soviet ICBM force in different directions. However,
even if Soviet fifth generation ICBMs keep pace with
U.S. ICBM modernization on the offensive end -- by
being able to threaten a substantial portion of U.S.
ICBMs even if they are proliferated, superhardened,
made mobile, dispersed, or actively defended -- Soviet
ICBMs nevertheless might themselves become more
threatened by the growing accuracy and numbers of U.S.
ICBM warheads.

The Soviet military leadership's first method for
offsetting the growing sophistication of forces
arrayed against it has been to reorganize in order to
bring the peacetime management of its armed forces
more in line with the necessities of wartime organiza-
tions. Marshal Ogarkov has led a drive to establish
the theater of military operations (TMO) as the basic
planning unit of Soviet operations and to create in
peacetime the commands for these wartime organization.
In addition, he has attempted to centralize operational control of all long range nuclear forces within
the general staff under the rubric Strategic Nuclear
Forces.[51] These changes should make the Soviet Union
able to switch to a wartime footing much more rapidly
than in the past. It should be noted that this drive
to improve the responsiveness of the Soviet command
structure probably dates back to the early 1970s when
the General Staff moved to increase its authority in
Soviet defense decisionmaking. The issue of vulnerability has been seized by Marshal Ogarkov as a con-
venient means for justifying the changes which he
desired and as a means for overcoming the resistance
of the services. Thus, it would be wrong to see the
institutional changes described above as just a
response to American weapons programs.

A second Soviet response to the very accurate
U.S. Peacekeeper and small ICBMs might be to increase
the readiness rates and to lower the response time of
threatened systems. The Soviet Union long has
threatened to launch its missile forces on warning of
an attack. Two current developments may improve its

ability to execute such a policy. First, the Soviet
Union continues to build large phased array radars
which can give its missile forces warning of an incoming attack. The Soviet Union now has six Pechora
class large phased-array radars operational or under
construction to add to an older network of eleven Hen-
house phased-array radars. In addition, the U.S.S.R.
maintains a launch detection satellite network and two
over-the-horizon radars capable of providing early
warning.[52] Second, the fifth generation of Soviet
ICBMs currently under development should have improved
reaction times, especially the new solid fueled missiles. Taken together these developments should
improve the Soviet Union's current ability to execute
a launch-on-warning strategy -- a strategy which they
reportedly have practiced under strict time constraints.[53]

Apparently, the Soviet leadership is not prepared
to count on such a launch-on-warning strategy alone.
During the introduction of new missile types, the
Soviets have increased the hardness of their ICBM
silos to a significant degree. This should provide
some additional survivability for missiles not
launched before an incoming attack arrived.

The Soviet ICBM force might gain an additional
measure of survivability with the deployment of mobile
systems. Two new ICBMs, the SS-X-24 and the SS-X-25,
currently being tested at Plesetsk, reportedly are
slated for both silo and mobile deployment. The
SS-X-25 will be based in launcher garages equipped
with sliding roofs and will be capable of off-road
mobility aboard massive wheeled transporter-
erector-launcher vehicles. The larger SS-X-24 will
probably first be deployed in silos with mobile
deployment following some years later.[54]

While the improved accuracy of U.S. systems like
the Peacekeeper and the SICBM probably encourages the
Soviets to continue the development of mobile systems,
the Soviet Union's interest in a mobile ICBM is of
long standing. In 1967 P.T. Astashenkov wrote, "Solid
fuel intercontinental and medium range missiles on
self-propelled launchers are a remarkable new
development. No other army in the world has a weapon
of similar effectiveness."[55] At the time, the Soviets
did not have an ICBM of this type. Astashenkov's
remark may have referred to the SS-13 missile just
entering service. The SS-13 may have originally been
intended as a mobile system, but was only deployed in
small numbers in silos. Apparently, it proved incapable of performing the mobile mission adequately.[56]
The SS-13 was not the only failure of the Soviet
mobile missile program. Neither the SS-14 nor SS-15
mobile IRBMs met performance standards or were

deployed in more than experimental numbers.[57] The
SS-16 ICBM was also designed for mobile deployment,
but may have run into problems as well. According to
the U.S. Defense Department report Soviet Military
Power:

> available information does not allow a conclusive
> judgement on whether the Soviets deployed the
> SS-16, but does indicate probable deployment.[58]

Difficulty with solid fuels plagued the Soviet mobile
missile program, but did not cause the Soviets to abandon their search for effective mobile systems. In
1977 the Soviet Union made something of a breakthrough
with initial deployment of the very capable mobile
SS-20 IRBM.

The long-standing Soviet effort to develop mobile
missiles indicates that systems such as the SS-X-24
and SS-X-25 are not direct responses to the Peacekeeper or to the SICBM, but are the logical outgrowth
of an indigenous Soviet program. Nevertheless, mobile
ICBMs should provide a partial antidote to the problems created for the Soviet Union by modernization of
the U.S. ICBM force. Mobile systems should be less
vulnerable to attack. Moreover, mobile ICBMs could be
withheld during the first phase of a war as part of a
strategic reserve force. Such systems would be far
less vulnerable than reloaded silos and stationary
soft-site launching pads, and they might provide
greater counterforce capability than surviving SLBMS.

The impact of highly accurate American ICBMs on
the survivability of Soviet nuclear systems may be
reduced further by the substantial effort which the
U.S.S.R.'s leadership is making to diversify its
nuclear forces. In particular, the development of the
Typhoon/SS-N-20 system and the follow-on SS-NX-23 SLBM
should enhance the capabilities of the relatively more
survivable submarine fleet.

While mobile missiles and improved sea-based
forces will make Soviet nuclear forces less vulnerable, they will do little to reduce the threat to political and military control centers and to the
U.S.S.R.'s urban-industrial base. To ensure leadership survivability and control over the armed forces,
the Soviets have long-standing programs to provide
multiple, hardened relocation centers and lines of
communication. Modernization of the U.S. ICBM force
will reinforce such efforts and may provide some
incentive to seek additional measures of protection
through ballistic missile defense.

The Soviet Union currently maintains the world's
only active BMD system, deployed around Moscow. This

system is being upgraded with the addition of a new
battle management radar at Pushkino and the addition
of new endoatmospheric and exoatmospheric interceptor
systems, the SH-08 and the SH-04.[59]

The Soviet Union has also taken steps to facilitate deployment of a much broader ABM systems in a
relatively short period of time. As discussed
earlier, the Soviets have built numerous large phased-
array radars for early warning. This radar network
might also provide the basis for a nation-wide ABM
system. To supplement this network, the Soviets have
developed "transportable" ABM radars to be deployed if
they decide to break out of the 1972 ABM Treaty. Moreover, the Soviets have tested surface-to-air missile
systems such as the SA-5, SA-10, and the SA-X-12 for
possible use against ballistic missiles.[60]

High level party encouragement has generated a
more dynamic R&D effort to exploit revolutionary BMD
technologies. The Soviets are making intense efforts
in the field of high energy lasers and particle beams.
Such technologies could be used either to destroy
satellites or for ballistic missile defense. Differences of opinion exist over how close the Soviets may
be to developing a working prototype based on the new
technologies.[61]

A BMD system would have definite advantages for
the Soviets. First, it could complicate, if not
totally deflect, a first strike. Second, it might
protect high-value political assets. Third, it might
provide a measure of area defense, especially if the
Soviets struck first. These advantages would have to
be weighed against the possible deployment by the
United States of a BMD system in response. Moreover,
BMD would require a major redirection of defense
investment and could entail the political cost of
breaking out of the 1972 ABM Treaty.

The Soviet Union's interest in BMD predates by
many years the development of highly accurate American
ICBMs and therefore should not be seen as a simple
reaction to a U.S. arms development initiative.
Indeed the Soviets first considered deployment of a
very crude ABM system around Leningrad in the early
1960s. The Soviet's interest in missile defense is
rooted in their belief that damage limitation, even in
a nuclear war, is an important objective. Given their
long-term interest in missile defense, Soviet efforts
in this area would probably continue even if U.S. programs like Peacekeeper and the SICBM were cancelled.

While critics of U.S. ICBM modernization may
argue that it will lead to dangerous Soviet
"reactions," it is likely that the Soviet Union will
continue with these "reactions" even in the absence of
Peacekeeper/SICBM deployment. Indeed most of these

"reactions" have roots which extend back well beyond the initiation of the current U.S. force modernization program. For the Soviets, accurate ICBMs are but one manifestation of dynamic changes in military technology which require a number of changes in military organization, force structure, and operational procedures.[62] Besides the rational reasons for continuing with their current defense programs, there are questions of domestic politics. The institutional momentum behind current Soviet program is such that U.S. action may not be able to affect the pattern of Soviet behavior.

Political Responses to ICBM Modernization

One of the principal goals of Soviet arms control policy is to impede the development and acquisition of advanced armaments by the United States. This policy takes two forms. First, the Soviets seek direct prohibitions against specific U.S. systems and technologies. Second, they try to create an environment which encourages domestic opposition in the United States to the U.S. programs in question. Both tactics have been used against U.S. ICBM modernization.

The Soviet Union attempted to prohibit Peacekeeper on several occasions during the SALT II negotiations. Soviet negotiators continually tried to limit exceptions to the ban on new ICBMs to single warhead missiles. This would have prohibited Peacekeeper while permitting the Soviets to deploy a solid fuel, single warhead replacement for the SS-11. In order to kill the Peacekeeper program, the Soviets eventually suggested that all new types of ICBMs be prohibited.[63]

The Soviets also argued that the construction of additional silos for either multiple aim point deployment or closely spaced basing violated the prohibition on new fixed launchers in Article 4, paragraph 1 of the SALT II Treaty.

The Soviets have also attempted to ban Peacekeeper and the SICBM as part of a proposed general freeze on nuclear force modernization. Such a freeze would leave the Soviets with a much newer ICBM force possessing greater hard-target capability than that of the United States. Moreover, the Soviets might argue later that a freeze did not prohibit deployment of their fifth generation ICBMs. They could argue that new missiles are really just versions of existing types and hence are not subject to a freeze on modernization. Reportedly, in fact, they have already used this argument to circumvent the limit on new ICBM types in the SALT II agreement. According to press reports, the Soviets have insisted that the SS-X-25 is not a new missile, but a variant of the SS-13.[64]

Soviet support for a "freeze" on new nuclear weapons is designed to reinforce opposition within the United States to the modernization of strategic forces. This has been a major goal of Soviet arms proposals since the 1950s. However, in the early 1980s, the Soviet leadership began to reassess the utility of arms control negotiations as a means for obtaining this end. Indeed, some Soviets suggested that START and INF were being used by the Reagan Administration to gain public support for systems which otherwise might not withstand the scrutiny of Congress.[65] This internal debate and reassessment of the utility of arms control negotiations did not take place all at once, and it was accompanied by considerable tension between Soviet elites.[66] After the initial deployment of Pershing II missiles began in the fall of 1983, however, the Soviet leadership apparently agreed that continued participation in the INF and START negotiations only helped the Reagan Administration to justify its arms programs to the public.

The Scowcroft Commission's initial success in selling Peacekeeper as an essential element of a new approach to arms control was further evidence to the Soviets that they could not slow down the United States strategic programs by continuing to negotiate. However, since many nuclear programs, including the Peacekeeper, had been sold to the public as useful to arms control, the Soviet leadership may have concluded that by refusing to negotiate they could undermine support for these systems. Where arms control negotiations served to increase public opposition to U.S. defense programs in the 1970s, the Soviets may now believe that their refusal to negotiate will achieve the same result. Either way, the goal remains the same -- to stimulate opposition to weapons such as Peacekeeper by manipulating arms control themes. Defeat of the Peacekeeper program in Congress could only reinforce the Soviet leadership's current belief that their refusal to negotiate is more productive than a return to the negotiating table.

Conclusions

The Soviet Union sees ICBM modernization as closely linked to recent changes in U.S. strategy. Modernization of the ICBM force is recognized by the U.S.S.R. as vital to the implementation of the new American nuclear strategy contained in PD 59 and developments under the Reagan Administration. While they study U.S. strategy carefully, the Soviets still interpret its provisions through the prism of their own tradition of strategic thought. In particular, they believe the Peacekeeper and the SICBM

would be most useful in a first strike attack.
Currently, the Soviet Union is implementing a
number of programs to reduce its vulnerability to
attack by very accurate systems like Peacekeeper and
the SICBM. These include measures: to bring the
peacetime and wartime organization of the Soviet
command structure closer together; to improve attack
assessment and launch-on-warning capabilities; to
increase the survivability of ICBMs through silo
hardening and mobility; to create a rapidly deployable
ABM system for defense of the leadership, nuclear
forces, and perhaps some of the population; to ensure
the survivability and control of the leadership even
without an ABM system; and to diversify Soviet nuclear
forces. In many cases, these programs predate
Peacekeeper and the SICBM and are likely to continue
even if the United States does not modernize its ICBM
force.
The Soviet Union also has conducted a political
campaign to prevent deployment of Peacekeeper. Soviet
negotiators tried to prohibit Peacekeeper in SALT II.
Similarly Soviet support for a freeze on nuclear force
modernization would prohibit both Peacekeeper and the
SICBM. Finally, the Soviets are attempting to undermine public support for systems such as Peacekeeper by
refusing to participate in the START talks. Since
Peacekeeper was sold by the Scowcroft Commission as
essential to arms control, the Soviets hope that by
scuttling the START talks they can undermine the
Administration's case in support of the system.

NOTES

1. Walter Slocombe, "The Countervailing Strategy,"
    International Security, Vol. 5, No. 4 (Spring
    1981), p. 19.

2. Ibid., p. 20; and Secretary of Defense Harold
    Brown,Department of Defense Annual Report, Fiscal
    Year 1982, (Washington, D.C.: GPO, 1981), p. 38.

3. Brown, Department of Defense Annual Report,
    Fiscal Year 1982, p. 40.

4. Slocombe, "The Countervailing Strategy" p. 21.

5. M.S. Mil'shtein, "Nekotorye kharakternye cherty
    sovremennoi voennoi doktriny SShA," SShA:
    ekonomika, politika, ideologiia, No. 5 (May
    1980), pp. 11-12.

6. M.S. Mil'shtein and L.S. Semeiko, "Problema nedopustimosti iadernogo konflikta," SShA: ekonomika, politika, ideologiia, No. 11 (November 1974), pp. 3-4.

7. Major-General R. Simonian, "V poiskakh 'novoi strategii'," Pravda, March 19, 1979.

8. L. Semeiko, "Stavka na potentsial pervogo udara," Krasnaia zvezda, August 8, 1980.

9. Ibid.

10. Mil'shtein, "Nekotorye kharakternye cherty sovremennoi voennoi doktriny SShA," p. 12.

11. Ibid., p. 13, emphasis added, and R.G. Bogdanov, M.A. Mil'shtein, and L.S. Semeiko, eds., SShA: voenno-strategicheskie kontseptsii; (Moscow: Nauka, 1980), p. 174.

12. A turning point in Soviet assessments came in the fall of 1982. See the speeches by Brezhnev, Chernenko, and Grishin reported in Pravda October 28 and 30, and November 6, 1982.

13. For example, see the TASS interview with Defense Minister Ustinov reported in Strategic Review, No. 1 (Winter 83), p. 85.

14. Brown, Department of Defense Annual Report, Fiscal Year 1982, p. 40.

15. Lev Semeiko, "Directive 59: Evolution or Qualitative Leap", New Times, No. 38 (September 1980), p. 5.

16. Simonian, "V poiskakh 'novoi strategii'."

17. Ibid.

18. Harriet Fast Scott, ed., Soviet Military Strategy (New York: Crane, Russak, 1975), p. 59.

19. D.F. Ustinov, "Protiv gonki vooruzhenii i ugrozy voiny," Pravda, July 25, 1982.

20. Marshal of the Soviet Union N. Ogarkov, "Vo imia mira i progressa," Izvestiia, May 9, 1982.

21. A. Arbatov, "Strategiia iadernogo bezrassudstva," Kommunist, No. 6 (June 1981), p. 106.

22. For example, see Harold Brown, _Department of Defense Annual Report, Fiscal Year 1982_, pp. 39-40.

23. Major General of the Engineering Technical Services I. Anureyev, "Determining the Correlation of Forces in Terms of Nuclear Weapons," _Voyennaya mysl'_, No. 6 (June 1967), (translated by the Foreign Broadcast Information Service), p. 38. (All _Voyennaya mysl'_ citations which follow were translated by FBIS.)

24. Scott, _Soviet Military Strategy_, p. 246.

25. M.M. Kir'ian, A.A. Babkov, A.N. Bazhenov, A.S. Begishev, _et.al._, _Voenno-tekhnicheskii progress i vooruzhennye sily SSSR_ (Moscow: Voenizdat, 1982), p. 314.

26. Colonel M. Shirokov, "The Question of Influence on the Military and Economic Potential of Warring States," _Voyennaya mysl'_, No. 4 (April 1968), p. 33.

27. Major General N. Vasendin and Col. N. Kuznetsov, "Modern Warfare and Surprise Attack," _Voyennaya mysl'_, No. 6 (June 1968), pp. 42-48.

28. Col. M. Shirokov, "Military Geography at the Present Stage," _Voyennaya mysl'_, No. 11 (November 1966), p. 63.

29. Ibid., p. 59.

30. Ibid., p. 59.

31. Anureyev, "Determining the Correlation of Forces in Terms of Nuclear Weapons," pp. 37-38.

32. Department of Defense, _Soviet Military Power_ (Washington, D.C.: USGPO, 1984), p.24.

33. Oleg Penkovskiy, _The Penkovskiy Papers_ (New York: Doubleday and Co., 1965), p. 257; and Colonel General N. Lomov, "O sovetskoi voennoi doktrine," _Kommunist vooruzhennykh sil_, No. 10 (May 1962), p. 19.

34. See the 1963 and 1968 editions of _Military Strategy_ in _Soviet Military Strategy_, p. 211 and N.V. Ogarkov, "Strategiia voennaia," _Sovetskaia voennaia entsiklopediia_, Vol. 7 (Moscow: Voenizdat, 1979), p. 564.

35. For example see S.P. Ivanov, Nachal'nyi period voiny (Moscow: Voenizdat, 1974).

36. Ogarkov, "Strategiia voennaia," p. 564.

37. This theme is developed in Nachal'nyi period voiny.

38. Scott, Soviet Military Strategy, p. 280.

39. See Clarence A. Robinson, Jr., "Soviet SALT Violations Feared," Aviation Week and Space Technology, Sept. 22, 1980, pp. 14-15; U.S. Congress, House, Committee on Appropriations, Subcommittee on the Dept. of Defense, Department of Defense Appropriations for 1983, Hearings, Part 4, 97th Cong., 2nd session, (Washington, D.C.: GPO, 1982), p. 544; and Soviet Military Power; p. 21.

40. "Otvet L. I. Brezhneva na vopros korrespondenta Pravdy," Pravda, October 21, 1981, p. 1.

41. A. Arbatov, "Strategiia iadernogo bezrassudstva," p. 106.

42. "Sveriaias' s Leninym, Deistvuia po-leninskoe: Doklad Tov. K. U. Chernenko," Pravda, April 23, 1981, p. 1.

43. A. Koldunov, "Moguchii shchit otchizny," Selskaia zhizn', February 23, 1982.

44. N.N. Azovtsev, V.I. Lenin i sovetskaia voennaia nauka, (Moscow Nauka, 1981), p. 66.

45. See Dan L. Strode and Rebecca V. Strode, "Diplomacy and Defense in Soviet National Security Policy," International Security, Vol. 8. No. 2 (Fall 1983), pp. 91-116.

46. Colonel G. Lukava, "Sovetskaia voennaia nauka i boevaia gotovnost'," Kommunist vooruzhennykh sil, No. 2 (January 1984), p. 32.

47. Clarence A. Robinson, pp. 14-15; "Soviets Stage Integrated Test of Weapons," Aviation Week and Space Technology, June 28, 1982, p. 20-21; and Michael J. Deane, Ilana Kass, and Andrew G. Perth, "The Soviet Command Structure in Transformation," Strategic Review, No. 2 (Spring 1984), pp. 55-70.

48. D.F Ustinov, "Otvesti ugrozu iadernoi voiny," Pravda July 12, 1982.

49. *Soviet Military Power*, p. 24.

50. On the effect of fractionation on accuracy see Bruce W. Bennett, *Uncertainty in ICBM Survivability*, P-6394 (Santa Monica: The Rand Corporation, 1979).

51. N.V. Ogarkov, *Vsegda v gotovnosti k zashchite otechestva* (Moscow: Voenizdat, 1982), p. 49 and *Krasnaia zvezda*, September 23, 1983.

52. *Soviet Military Power*, pp. 32-33.

53. Ibid., p. 20.

54. Ibid., p. 24.

55. P.T. Astashenkov, *Sovetskie raketnye voiska*, (Moscow: Voenizdat, 1967), pp. 101-102.

56. Robert P. Berman and John C. Baker, *Soviet Strategic Forces* (Washington, D.C.: The Brookings Institution), pp. 120-121.

57. Ibid., p. 111.

58. *Soviet Military Power*, p. 24.

59. Ibid, p. 33; and Robinson, "Soviet SALT Violations Feared," p. 15.

60. *Soviet Military Power*, p. 34.

61. "Soviet Build directed Energy Weapon," *Aviation Week and Space Technology*, July 28, 1980, pp. 47-50.

62. Ogarkov, *Vsegda v gotovnosti k zashchite otechestva*, pp. 313-44.

63. Strobe Talbott, *Endgame: The Inside Story of SALT II* (New York: Harper and Row, 1979), pp. 165-167.

64. *Washington Post*, August 13, 1983.

65. Ibid., November 3, 1982.

66. Strode and Strode "Diplomacy and Defense in Soviet National Security Policy," pp. 108-109.

9
Peacekeeper's Role in Deterrence: Concluding Comments

*Keith B. Payne, Rebecca V. Strode,
and Michael Ennis*

Introduction

When the country was still being developed, U.S.
Marshalls maintained order and deterred lawlessness on
the Western frontier with the aid of a Colt 45. Over
the years this weapon, the Peacemaker, became the
symbol of the rule of law. It is not surprising,
therefore, that President Reagan chose to link the
function of the MX missile to that of the Colt 45.
Where the Peacemaker brought peace to a lawless
frontier, the Peacekeeper is intended to help maintain
peace in an unstable world.

A Necessary Response to the Soviet Nuclear Buildup

It is sometimes argued that deployment of the
Peacekeeper Intercontinental Ballistic Missile (ICBM)
would escalate the arms race by prompting the Soviet
Union to deploy even more nuclear forces. This argument seriously misrepresents the current strategic
balance and inaccurately portrays as a U.S. initiative
what is in fact a critical response to the Soviet
Union's on-going missile buildup. Deployment of 100
Peacekeeper ICBMs represents an important, but nevertheless modest, response to the large scale missile
deployments which the Soviet Union began in the 1970s
and is continuing at the present time.

The U.S.S.R. has already deployed 308 SS-18 heavy
ICBMs and 360 somewhat smaller SS-19 ICBMs, for a
total of 668 currently deployed ICBMs with throwweight
equal to or greater than the Peacekeeper's. These
Soviet missiles are highly accurate and are deployed
in silos significantly harder than U.S. silos. They
thus constitute a formidable threat to the United
States, and one which can not be countered with the
currently deployed U.S. strategic missiles. Moreover,
the U.S.S.R. is already developing a whole new genera-

tion of ICBMs, including the SS-X-24 (which will achieve initial operational capability next year and will be even more accurate than the SS-18 and SS-19), and the SS-X-25 (apparently designed for mobile deployment).[1] Thus, deployment of 100 Peacekeeper ICBMs is essential if the United States is to redress the current and growing imbalance in prompt, countermilitary potential. Since the Soviet Union has already deployed more than six times the number of Peacekeeper-class ICBMs than the United States plans to deploy, Peacekeeper deployment can hardly be considered an escalatory action. Rather, it is a necessary and measured response to a Soviet nuclear buildup.

The strategic environment in which U.S. strategic forces must fulfill their deterrence mission has changed dramatically since the deployment of Minuteman III in the early 1970s. In particular, the Soviet Union possesses a significant advantage in counterforce capability. This enables the U.S.S.R. to pose a far greater threat to U.S. deterrent forces than the U.S. currently is able to pose to the Soviet Union. If Western deterrence is to remain strong, the United States must redress this imbalance by developing the capability to hold at risk high-value Soviet political and military assets. Such a capability would impress upon the Soviet leadership that it can gain no advantage by initiating a nuclear strike.

The Peacekeeper is a vital part of the necessary U.S. modernization effort. Minuteman II and III ICBMs do not possess the necessary capability to threaten comprehensively Soviet hardened military and political assets. Peacekeeper provides this capability. This will provide the increased capability required to cope with the growing Soviet hard-target structure. Peacekeeper will be almost twice as accurate as the Minuteman III, assuring that its warheads will be able to more effectively threaten hardened Soviet targets.

In brief, Peacekeeper provides the capabilities which are essential if the United States is to maintain a credible deterrent through the 1990s.

How Peacekeeper Contributes to Stability

A credible and reliable deterrent must threaten those values that the Soviet leadership holds dear. The Soviet leadership must be convinced that an attack on the U.S. or its allies would entail unacceptable retaliatory costs. As the Scowcroft Commission observed, the Soviet leadership values most highly its instruments of political and military power, and those facilities should be threatened to deter Soviet attack on the U.S. and its allies:

In order to deter such Soviet threats we must be
able to put at risk those types of Soviet
targets--including hardened ones such as military
command bunkers and facilities, missile silos,
nuclear weapons and other storage, and the rest--
which the Soviet leaders have given every indication by their actions they value most, and which
constitute their tools of control and power.[2]

The notion that the U.S. needs the capability to
threaten Soviet hardened military and political control assets, a so-called "prompt hard-target" capability, is not unique to the Reagan Admininstration.
Harold Brown, Secretary of Defense during the Carter
Administration, maintains that U.S. deterrence requirements should include 1,000-2,000 prompt hard-target
capable warheads. Such a force is necessary in order
to provide an "effective ability to attack a Soviet
political leadership that is being shuttled among
hardened command posts."[3]
Over the last decade the Soviet Union has
expanded the number of political and military facilities and has "superhardened" many against nuclear
attack. Because of this, the unique capabilities of
the Peacekeeper ICBM, including promptness and the
accuracy required to attack a hard-target, are now
necessary to hold those Soviet instruments of military
and political power at risk and thereby sustain the
U.S. deterrent against nuclear war. Peacekeeper will
ensure that the Soviet Union continues to be discour-
aged from contemplating attack against the U.S. and
its allies.
Opponents of Peacekeeper have criticized the
missile as as destabilizing, "first strike" weapon.
However, the Reagan Administration cut in half the
number of missiles that had been scheduled for deployment by the Carter Administration, from 200 to 100.
These would provide 1,000 or fewer warheads. Such a
force simply would be incapable of threatening the
entire Soviet ICBM force of 1400 missiles. Even when
complemented by Minuteman III, a comprehensive high
confidence threat to the Soviet Strategic Rocket
Forces could not be mounted. Moreover, even if U.S.
ICBMs one day become theoretically capable of posing a
significant threat to Soviet ICBMs, they would still
be unable to negate Soviet submarine ballistic missile
forces or long-range bombers and cruise missiles.
Peacekeeper is neither needed nor desired as a first
strike weapon but, as Harold Brown observed, it is
needed to threaten Soviet hardened military and political leadership facilities for the purpose of maintaining deterrence.

Opponents of Peacekeeper also have criticized the missile as placing a "hair trigger" on the U.S. nuclear forces. The line of reasoning is that because Peacekeeper will be deployed in theoretically vulnerable missile silos the Soviet Union will be tempted to launch an attack against those silos, and the U.S. will be forced to launch first in order to ensure the survival of the ICBMs. Actually, the margin of doubt concerning the Soviet capability to launch a comprehensive and effective attack on U.S. ICBMs is very large. The Soviet leadership would confront tremendous uncertainties in contemplating such an attack-- uncertainties that serve to enhance deterrence. The Soviet Union could not be confident that it could effectively attack Peacekeeper, or that most Peacekeeper missiles would not be launched out from under the Soviet attack following the initial detonation of Soviet warheads. Deployment of Peacekeeper in silos will be adequate to ensure that the U.S. will have a credible capacity to hold at risk those targets critical to the maintenance of deterrence.

Peacekeeper and the Strategic Triad

The Scowcroft Commission emphasized that the United States deploys a "Triad" of strategic forces, including ICBMs, strategic bombers, and submarine-launched ballistic missiles (SLBMs). The differing basing modes of the three legs of the Triad ensure that if the Soviet Union develops a capability to threaten any individual leg, it will not have achieved a first strike capability. Opponents of Peacekeeper often discuss the missile out of the context of the Triad, as if the U.S. and the U.S.S.R were simply two ICBM fields. Such neglect of the actual strategic context misrepresents the characteristics of the Triad and the role Peacekeeper will play within the Triad. It is the Triad, not any single strategic weapon, which ensures that the Soviet Union cannot attain a comprehensive first strike capability--thereby discouraging Soviet nuclear attack. Within this strategic Triad, Peacekeeper will satisfy the current requirement for increased prompt, hard-target capability.

In short, the unique capabilities of Peacekeeper are essential to deterrence. Deployment of 100 Peacekeeper ICBMs will not constitute a first strike capability, nor will it encourage a Soviet preemptive attack. Rather, it will enhance deterrence by enabling the United States to provide a credible threat appropriate to the particular values of the Soviet leadership, i.e., a capability to hold at risk those hardened military and political control facilities identified as most important to that

5.   A version of this chapter first appeared in <u>Issue</u>
     <u>Bulletin</u> No. 106, May 31, 1984, the Heritage
     Foundation, Washington, D.C.

Acronyms and Abbreviations

ABM	anti-ballistic missiles
ACDA	Arms Control and Disarmament Agency
ALCM	air-launched cruise missiles
ASAT	anti-satellite weapon
ASW	anti-submarine warfare
BMD	ballistic missile defense
C^3	command control and communications
CD	U.N. Committee on Disarmament
CONUS	continental United States
CPSU	Communist Party of the Soviet Union
CSB	closely spaced basing
DARPA	Defense Advanced Research Project Agency
DDR&E	Director Defense Research and Engineering
DEW	direct energy weapon
DIA	Defense Intelligence Agency
DoD	Department of Defense
DSAT(s)	defense for satellite(s)
DSMP	defense meteorological program
ECM	electronic countermeasures
ENMOD	Environmental Modification Techniques
ERCS	emergency rocket communication system
ER	enhanced radiation
FY	fiscal year
FYP	five year plan
GLCM(s)	ground-launched cruise missile(s)
H^O	hydrogen atoms
HOE	homing overlay experiment
ICBM(s)	intercontinental ballistic missile(s)
INF	intermediate range nuclear forces

IOC	initial operation capability
LoAD(s)	low altitutde defense system
LODE	large optics deomonstration experiment
LOW	launch-on-warning
LWIR	long-wave infrared
MAD	mutual assured destruction
MARV(s)	maneuverable reentry vehicles(s)
MEV	million elechon volts
MIRV(s)	multiple independently targetable reentry vehicles(s)
MPS	multiple protective shelter
MT	megaton
MX	developmental American ICBM
NATO	North Atlantic Treaty Organization
NCA	National Command Authority
NNK	non-nuclear kill
NORAD	North American Air Defense
PD	Presidential Directive
PRO	anti-missile defense (Soviet)
PVO	anti-air defense (Soviet)
R&D	research and development
RDT&E	research, development testing and engineering
SAC	Strategic Air Command
SALT	strategic arms limitation talks
SAM(s)	surface-to-air missile(s)
SBL(s)	space based laser(s)
SCC	Standing Consultative Commission
SIOP	Single Integrated Operation Plan
SLBM	submarine launched ballistic missile
SRAM	short-range attack missile
SRF	strategic rocket force (Soviet)
SSBN	nuclear ballistic missile carrying submarine
START	Strategic Arms Reduction Talks

About the Contributors

Michael Ennis, a member of the National Institute for Public Policy professional staff, specializes in international relations and national security analysis. Mr. Ennis holds an M.A. in international relations from Catholic University where he currently is a doctoral candidate. Prior to joining the National Institute, Mr. Ennis served as a visiting fellow at the Foreign Policy Research Institute and was also an instructor of political science at Randolph- Macon College.

Colin Gray, president of the National Institute for Public Policy, is a strategist with a professional background in the study of U.S., Soviet and NATO defense policies. Dr. Gray studied at the University of Manchester and Oxford University in the United Kingdom and has taught at the universities of Lancaster, York, British Columbia and Georgetown University, Washington, D.C. Dr. Gray's most recent books include: American Military Space Policy (1983); Strategic Studies & Public Policy: The American Experience (1982); Strategic Studies: A Critical Assessment (1982); The MX ICBM and National Security (1981).

Keith Payne, executive vice president of the National Institute for Public Policy, is a political scientist specializing in areas of U.S. and Soviet defense and foreign policy. Dr. Payne is a graduate of the University of California at Berkeley, and the University of Southern California where he received a Ph.D. in international relations. Dr. Payne is co-author of Nuclear Strategy: Flexibility and Stability (1979), the author of Nuclear Deterrence in U.S.-Soviet Relations (Westview, 1982), and editor and contributor to Laser Weapons in Space (Westview Press, 1983) and The Nuclear Freeze Controversy (1984).

Barry Schneider is a senior defense analyst at the National Institute for Public Policy, specializing in international politics, defense and arms control. Dr. Schneider formerly worked on arms control and military affairs for various members of Congress and was a foreign affairs officer in the U.S. Arms Control and Disarmament Agency from 1977-1980. He has a Ph.D. from Columbia University and has taught at Wabash, Purdue, Indiana, Maryland, Georgetown and American. Dr. Schneider is the co-editor of Current Issues in U.S. Defense Policy (1976) and has published more than fifty articles and professional papers.

Blair Stewart is a senior scientist with JAYCOR, a research and development firm. A former U.S. Air Force officer, he has an extensive background in the development of ICBM weapons systems, including Minuteman II, Minuteman III and MX. Before leaving the Air Force, he served as a staff officer in the Office of the Special Assistant for MX Matters, Headquarters U.S. Air Force. He is a 1968 graduate of the United States Air Force Academy and holds a masters degree in industrial systems engineering.

Dan Strode is a Ph.D. candidate in Harvard University's Department of Government. He has been an analyst of Soviet defense policy for the National Institute for Public Policy and has worked for the Department of Commerce on Soviet Research Development Practices. His articles have appeared in International Security, Orbis, and Soviet Union/Union Soviétique.

Rebecca V. Strode is a Soviet defense and foreign policy analyst at the National Institute for Public Policy. She is co-author of Areas of Challenge for Soviet Foreign Policy in the 1980s (Indiana University Press, 1984). She has also published articles on Soviet national security and foreign policy in a variety of journals. Ms. Strode received an M.A. in Soviet Studies from Harvard University in 1979.

Index